Earthquake Vapor Model & Precise Prediction

Zhonghao Shou & Yan Fang

Library of Congress Cataloging-in-Publication Data

Zhonghao Shou & Yan Fang

ISBN: 978-0-9975730-0-8
LCCN: 2016911717

Cover Designer: Yan Fang
Editor: Wenying Shou
Publisher: Earthquake Prediction Center, Seattle, WA

earthquake.prediction@gmail.com
fangyan@ucla.edu
http://eqclouds.wix.com/predictions

Dedicated to those who suffered in earthquakes

Preface

Earthquakes are thought to defy predictions. Indeed, numerous prediction methods have been attempted by generations of geologists. These methods are based on correlations of phenomena (such as animal behavior, tidal behavior, radon generation, ground shift, electromagnetism, foreshocks) with earthquakes. However, none of these methods have yielded useful predictions, because they are based on correlations instead of mechanisms. That is, an impending earthquake may or may not lead to these putative precursors, and phenomena other than earthquakes may also cause these phenomena.

In this book, we will discuss an important discovery– earthquake vapor theory – that can be used to generate precise and accurate earthquake predictions. This theory proposes that when a huge rock is stressed by external (including man-made) forces, its weak parts break first. This induces crevices into which ground water percolates. Water expansion, contraction, friction and erosion further weaken the rock. Friction during ground movement heats the ground water and eventually generates vapor at high temperature and pressure. This vapor erupts from an impending hypocenter to the surface by the crevices and a nozzle, and rises up to form an "earthquake cloud" when encountering cold temperature at a higher altitude or a geoeruption (geothermal eruption) as its heat dissipates a preexisting cloud.

Earthquake clouds and geoeruptions can be differentiated from other geological or weather phenomena because they are vapor-based, suddenly appear, and have a fixed source, high temperature and high pressure. The location of eruption predicts the epicenter, the vapor amount predicts the magnitude, and the time after a complete eruption empirically predicts the time of the earthquake within days.

Using this theory, Shou predicted 63 independent earthquakes to the United States Geological Survey from 1994 to 2001. Each prediction has three definite windows of time, area and magnitude. More than 60% of these predictions are all correct. An evaluation on this set shows that random guesses will have only a chance of 0.001 to achieve the same level of success. Shou also made over 1500 predictions to the public from 1999 to 2007. The Bam cloud is representative. It appeared suddenly from and insisted at a nozzle in Bam, Iran for 26 hours on December 20~21, 2003 when surface temperature increased $5^{\circ}C$ downwind. The Bam prediction on December 25 is excellent. An M6.8 earthquake hit Bam exactly on December 26 and was the only one in the predicted area and magnitude in the record of about 3000 years. The United Nations published my paper "Bam Earthquake Prediction & Space Technology" in its 2004 yearbook (*Seminars of the United Nations Programme on Space Applications* **16**), and shared this book to its all member states in Vienna in early 2005. The United States Patent and Trademark Department published "Method of precise earthquake prediction and prevention of mysterious air and sea accidents" (*Patent US 08068985B*) in 2011.

Despite the promise of the Earthquake Vapor Theory, I have received no fund to overcome technical difficulties associated with satellite data, temperature data, and earthquake data. I hope that this book will help the mankind to solve these solvable problems and be able to predict all devastating earthquakes.

I acknowledge LinYing Fang, WenYing Shou, Abdolreza Ansari Amoli, Andy M Han, Darrell Harrington, ChengNan Xu, Sri Ram, Liz Thomas, JianJun Xia, XiaoKun Chen, ShuiZhen Ye, Guohai Ding. JuanJuan Shao, the Lee & Hayes Office, and the Han Office for support or help. I thank international sources, media, scholars, and people for data, reports, information, inspiration, and donation.

Contents

Chapter 3   Evaluation of earthquake predictions

Chapter 4 Critiques of the Plate Theory

Chapter 1   Earthquake Vapor Model

The earthquake vapor model was first proposed by Shou (1999). He hypothesized that under external stress, the weak parts of rocks break first. These fractured rocks allow water penetration. When rocks and groundwater move against each other, friction generates a tremendous amount of heat. Heat transforms groundwater to vapor of high temperature and pressure, which erupts through crevices to surface. The vapor rises up and condenses into a cloud after encountering colder air. The cloud tail points toward the epicenter; the size of the cloud indicates the magnitude of the impending earthquake; and the timing of earthquake was empirically determined to be within 49 days after the first appearance of the cloud. Shou named these clouds as "earthquake clouds."

Earthquake vapor model was further developed (Harrington and Shou, 2005) to include a new atmospheric phenomenon called geothermal eruption or "geoeruption" as an additional precursor. Geoeruption emerges as a sudden localized atmospheric heating or disappearance of cloud. Further, the warm region persists despite the presence of moving clouds overlapping or in the vicinity of the region. The paper extended the upper bound of time window to 103 days based on statistics on a larger sample size. However, it was still difficult to pinpoint an impending epicenter to a small area in warm surrounding, and to narrow a time window from months into a week.

Shou (2011) proposed solutions to both problems in Patent US 8,068,985 "Method of precise earthquake prediction and prevention of mysterious air and sea accidents." Below, we describe the earthquake vapor model and how it may be used to precisely predict earthquakes.

For the entire book, unless otherwise specified (such as "LT" for Local time), all time is in UTC (Coordinated Universal Time), all temperature values are in Celsius (^{o}C), and all pressure values are in atmosphere pressure (atm). Coordinate follows the custom of the United States Geological Survey (USGS): latitude before longitude, and positive for north and east, while negative for south and west. For example, Bam (28.99, 58.29) is at latitude 28.99N and longitude 58.29E; Rio de Janerio (-23, -43.4) is at latitude 23S and longitude 43.4W. Several samples of Shou's predictions, including the Bam earthquake prediction, witnessed by scientist or USGS are exhibited by the appendix against rumor.

1.1 Water penetration

The first critical element of the Earthquake Vapor Model is the underground water postulated to exist kilometers deep underneath the surface, even below hypocenters of earthquakes. We describe potential mechanisms of water penetration into deep underground.

When a huge rock is stressed by external forces, its weak parts break first. Table 1 shows all large quakes (M≥6.0) in Southern California from 1980 to 2012. For each large quake, numerous smaller shocks occurred earlier and within 10 km to the epicenter. These smaller shocks are expected to generate small crevices in a manner similar to how large earthquakes generate big rifts. These small crevices in turn reduce the cohesion of the rock.

Table1. All big earthquakes (M≥6) in Southern California from 1980 to 2012 and their nearby prior shocks

No	Date	Time	Latitude	Longitude	Mag≥6	Depth km	Prior Shocks in 10km Total	Deeper
1	19830502	23:42	36.23	-120.32	6.1	10.2	22	1
2	19871124	1:54	33.09	-115.80	6.2	10.8	138	10
3	19871124	13:15	33.01	-115.86	6.6	11.2	558	33
4	19920423	4:50	33.96	-116.32	6.1	12.3	1602	14
5	19920628	11:57	34.20	-116.44	7.3	1.0	520	461
6	19920628	15:05	34.20	-116.83	6.3	5.4	345	256
7	19940117	12:30	34.21	-118.54	6.7	18.4	79	5
8	19991016	9:46	34.59	-116.28	7.1	0.0	430	373
9	20031222	19:15	35.70	-121.11	6.5	7.0	37	7
10	20040928	17:15	35.81	-120.38	6.0	5.5	90	79

Note: Data are from the Southern California Earthquake Data Center (SCEDC@1) of the United States Geological Survey (USGS). The number of smaller prior earthquakes was since 1980, except in the case of No.1 which was since 1932 (Column "Total"). The last column indicates the number of smaller prior shocks deeper than the hypocenter of the large earthquake. Lat.: Latitude. Lon.: Longitude. Mag: Magnitude. No.7, 8 and 10 are the Northridge earthquake, the Hector Mine earthquake, and the 2004 Parkfield earthquake respectively.
@1http://www.data.scec.org/ftp/catalogs/SCSN/

Next, underground water percolates into the crevices. Its expansion, contraction, corrosion, and fluid friction further reduce rock cohesion. Fig.1 reveals a vertical distribution of shocks surrounding the Northridge hypocenter and prior to the Northridge earthquake. A shock 0.2 km below the hypocenter occurred on Mar.21, 1991 (Fig.1, A). Together with other smaller shocks, this suggests that ground water could have penetrated through and beyond hypocenter.

Diamond formation gives additional evidence of water penetration. On the one hand, chemical theory and manufacture of artificial diamonds show that the formation of diamond requires a high pressure of over 45 kilobars and a temperature of over 1000°C, which indicates a depth exceeding 150 kilometers under natural conditions. On the other hand, diamonds could exit from a depth of over 150 kilometers to the surface through kimberlite pipes (Cox,1978), which could conceivably also allow water trafficking.

Besides earthquakes, many processes can also produce crevices. Examples include weather processes (e.g. temperature change, wind, rain, snow, ice,

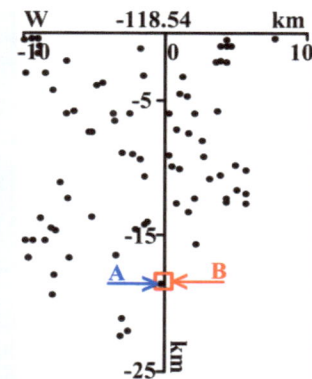

Fig.1 Vertical distribution of smaller prior shocks surrounding the Northridge hypocenter

This East-West vertical cross-section is through the Jan.17, 1994 Northridge epicenter (34.21, -118.54). The red box marked as B is the hypocenter at depth 18.4 km. All prior shocks within 10 km to the epicenter since Jan.1, 1980 were smaller than the big shock and projected and plotted as dots. A shock arrowed with A was 0.2 km below B. The data are from the SCEDC@1.

flood, and drought); geophysical activities (e.g. volcano, landslide, land rise, sinkhole, ocean currents, changes in earth-sun gravitational attraction or earth rotation, sunspot explosion, and meteors); geochemical activities (e.g. the formation of gas, oil, coal, and limestone caves, and decomposition into carbon dioxide); and human activities (e.g. drilling, explosion, mining, transportation, and dam construction).

The relationship between water in the rocks and earthquakes is reinforced by Bolt, (1978). Bolt reported that the USGS performed an experiment at the Rangely Oil Field in Western Colorado in 1969. In the experiment, water was regularly injected into or pumped out of the oil wells. Excellent correlation was found "between the quantity of fluid injected and the local earthquake activity." Bolt further proposed, "If there were no water in the rocks, there would be no tectonic earthquakes."

1.2 Sources of Friction

The Earthquake Vapor Model posits that friction vaporizes underground water which then forms precursors. Many processes can generate friction within and between rocks and underground water. Some occur with regularity. The Sun and the Moon are known to cause solid tides, ocean tides, and atmospheric tides. Thus, they should also cause underground water tides in crevices. Solid tides and underground water tides should induce friction during movements. Changes in the rate of earth rotation, evidenced by wobbling the pole and lengthening of a day at a rate of about 1~2 ms every century (Lambeck, 1980), can also induce friction between earth masses. .

Other friction-generating physical processes are less regular. Weather processes including precipitation, temperature change, hurricane, tornado, ocean current, drought, and flood can all cause friction through redistributing earth mass. For example, a heavy rainfall in Shenyang, China from Jul.31 through Aug.1,1975 had caused a big change on a tilt meter (Haicheng Earthquake Study Delegation, 1977). Earthquake and volcano can also cause frictions. Large earthquakes were reported to change the rotation of earth (Hopkin, 2004), which in turn causes friction. Earthquakes cause rupture, tremor, landslide and tsunami that also directly induce friction within and between rocks and underground water.

Human activities, such as drilling, explosion, mining, transportation, and dam construction change mass distribution. A team of scientists of the National Aeronautics and Space Administration (NASA) and the USGS found that the surface of Los Angeles varied in 1996~2001. These ups and downs in surface turned out to result from water companies storing and drawing underground drinking water (Clarke, 2001). These human activities can also induce friction within and between rocks and underground water.

Friction, resulting from natural or man-made processes, causes heat. In an open space, such as land and ocean surface, the heat is released into atmosphere. In a closed space, e.g. underground water in crevices of rocks, heat from the friction within and between moving rocks and underground water accumulates gradually. This would induce a high temperature.

1.3 High temperature

A prediction of the Earthquake Vapor Model is high temperature being associated with earthquakes. This is indeed the case. For example, part of the ice in the shady area of a frozen reservoir had melted away during a very cold winter before the 1975 Haicheng earthquake (Yang, 1982). On Dec.20 2003, six days before the M6.8 Bam, Iran earthquake, a pulse of surface temperature increase from 12°C to 24°C was recorded by the Kerman airport (30.2, 57) near Bam at 16:20 or LT 20:20 when the Bam cloud was erupting in a cold evening (Shou et al., 2010). Similarly, a surface temperature as high as 141°C was reported in Kerman at 14:20 or LT 18:20 on Dec.15, 2004 (Shou et al., 2010) when the vapor was erupting preceding the M6.5 Kerman earthquake and its partners (Shou, 2006a).

The M7.8 Tangshan earthquake on Jul.27, 1976 gave great examples of high temperature prior to, during, and after the quake (Shi et al., 1980). Very hot erupting matter had burnt a man during the quake. Prior to the earthquake, a 7.8-meter-deep dry well 150 km away from Tangshan (at Wan-Quan-Zhuang, Beijing) had been sounding like a steam whistle. A lot of gas had been erupting from it from Jul.26 until five hours before the Tangshan earthquake. After the quake, the dry well began whistling again. The plume of gas reached 2.5 meters high, and the velocity of the gas was 38 m/s. The sound reached 94 dB, and could be heard 200 meters away. Nine hours later, the M7.1 Luanxian quake struck near the original M7.8 quake. The gas plume appeared again before each of the two M6 shocks on Aug.8 and 9. The gas was analyzed as having 12.9% carbon dioxide (CO_2), compared to the CO_2 concentration of 0.04% in normal atmosphere (Shi et al., 1980). This analysis suggests that limestone was decomposing into carbon-dioxide and calcium oxide. Because decomposing temperatures of pure carbonate calcium and carbonate magnesium are 848°C and 360°C respectively, the hypocenters should have had a temperature of 360~848°C.

Using microscopy to investigate the structure of erupted rocks near epicenters, scientists found melt rocks, crystallite, and sudden change of chemical compositions (Koch and Masch, 1992; Maddock, 1992; Magloughlin, 1992; O'Hara, 1992; Spray, 1992; Swanson, 1992; Techmer et al., 1992). Measurements using friction welding machine, thermal dyes, and silicon dioxide glass composition analysis further revealed that the melting temperatures for these rocks ranged from 300°C to 1,520°C (Bowen and Aurousseau, 1923; Killick, 1990; Maddock, 1983; Passchier, 1982; Sibson, 1975; Spray, 1987; Tuefel and Logan, 1978; Wenk and Weiss, 1982; Winkler, 1979). Thus, the impending hypocenters could reach a temperature range of at least 300~1,520°C. Since water boiled at 300°C and 86 atm (Haas, 1971), a temperature of 300~1,520°C can make underground water boil.

1.4 High pressure

The Earthquake Vapor Model hypothesizes that vapor generated by high temperature will build up to high pressure in closed underground. Indeed, high pressure often accompanies earthquakes. Petroleum erupted about 20 meters high from a close well eleven days before the M7.8 Tangshan earthquake (Shi et al., 1980). Water spouts erupted to as high as 115 feet above the valley floor at an estimated 400 cubic feet per second during the M7.3 Borah Peak, Idaho earthquake on Oct.28, 1983 (Lane and Waag, 1985). Rocks erupted from the ground during the M7.7 Chi-Chi, Taiwan earthquake on Sep.20, 1999, forming a hole of 4-meter-wide and 40-meter-deep (Huang et al., 2003). Fig.2 reveals damaged ceiling due to the eruption of steam from underneath it during the Tangshan earthquake.

Before the M7.3 Haicheng earthquake on Feb.4, 1975, the production of oil well Re10-6 increased from pumped 4~17 tons/day in Nov. 1974 to spontaneous production of 80~90 tons/day in the early Dec. 1974; the pressure of oil at the bottom of well Xing 5 increased from 116.8 atm. on Oct.8 to 137 atm. on Oct.11, 1974 (Wu and Liu, 1983). Before the Tangshan earthquake on Jul.27, 1976, the oil production of well No.8 increased by six-fold compared to that in April, and the pressure rose by 20~50 atm in June (Zhang and Zhao, 1983). Five hours before the Tangshan earthquake, air erupted from a dry well in Beijing, 165-km from epicenter (Shi et al., 1980). The vapor of the M9 Sumatra quake in Indian Ocean was deduced to have a pressure of at least 151MP (1532 atm) to rise from the seabed of 16.1km-deep to sea-surface (Shou, 2006b).

7－17 丰润县宣庄公社一平房内喷沙冒水，冲破了房屋顶棚(10度区)。
In the area of intensity 10, sand boiling and water spouting occurred to a house in Yizhuang Commune in Fengnan County and spoiled the ceiling.

Fig.2: Damage of ceiling due to vapor eruption in the Tangshan earthquake. Photograph from (China Academy of Building Research, 1986).

1.5 Vapor eruption

Earthquake vapor has high temperature and high pressure. Thus, once the pressure of vapor surpasses the resistance of a main crevice, vapor can suddenly erupt from the hypocenter to the nozzle of the surface through the main crevice (Fig.3). Sometimes, uprising vapor encounters cold air and condenses to form a cloud called "earthquake cloud" (Shou, 1999). Sometimes, hot vapor encounters a pre-existing cloud and melts part of the cloud away, a phenomenon termed "geoeruption" (Harrington and Shou, 2005). In both cases, the vapor contains not only gaseous water, but also tiny water droplets of different sizes. The droplets are warm, and sometimes can form "earthquake fog" and heat up wherever the fog lands. Simultaneously, a small fraction of vapor can slowly escape through small crevices to the surface to form a warm band between the epicenter and the main vapor nozzle (Fig.3). After complete eruption, an earthquake follows soon. After an incomplete eruption, remaining water and vapor erupts again while accumulating enough

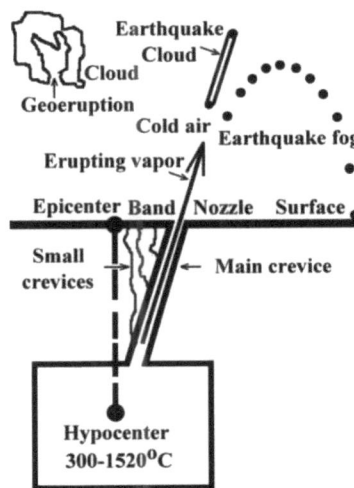

Fig.3. Earthquake vapor schematic diagram

energy, and then an earthquake follows.

1.6 Dehydration

Shou (2011) found a difference between a complete vapor eruption (Fig.4a) and an incomplete vapor eruption (Fig. 4b). After a complete eruption, crevices of a rock or an impending hypocenter becomes empty. Namely, almost no vapor was left inside to bear outside pressure. Moreover, temperature of the rock reaches a threshold where the yield strength of the rock reduces sharply, which is called "dehydration" (Kirby and McCormick, 1990). After a complete eruption, which happens about 10% of time, an earthquake follows within 3 days. The remaining 90% of eruptions are incomplete, and a complete eruption will follow within about 112 days. In this case, an earthquake will follow within 3 days after the second, complete eruption.

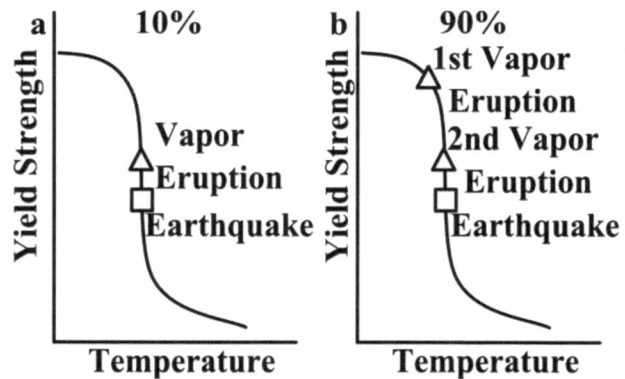

Fig.4 Dehydration

(**a**) Shou predicted the above characteristic curve of rocks in Apr. 1997. With help of a librarian of Caltech, he found measurements from the Practical Handbook of Physical Properties of Rocks and Minerals (Kirby and McCormick, 1990) which supported his prediction. This character was denoted "Dehydration" that perfectly matched his hypothesis: an earthquake happened when the yield strength of a rock reduced sharply after vapor eruption.

(**b**) Shou, (2006a) found that 10% earthquakes happened within 3 days after a vapor eruption. For the other 90% earthquakes, Shou (2011) found that the first eruption was incomplete and was followed by a second, complete eruption and a second temperature peak. Three days after the second round of eruption, an earthquake occurred. The longest delay from the first vapor eruption to an earthquake was about 112 days.

1.7 Examples of earthquakes and their precursors

1.7.1 Northridge Earthquake

Let's analyze how precursors of several big earthquakes fit the Earthquake Vapor Model. The first example is for the M6.7 Northridge earthquake (34.21, -118.54) on Jan.17, 1994. At LT 7:15 am or UTC 15:15 on Jan.8, 1994, a cloud rose up like a launching rocket in the northwest sky of Pasadena (34.14, -118.14), Southern California (Fig.5a). Meteorology cannot explain the sudden, vertical appearance of this cloud.

From this cloud, Shou predicted on Jan.15, 1994 to the USGS that an earthquake of magnitude greater than 6 would strike within 100 km northwest from Pasadena within 25 days of the appearance of the cloud. Unfortunately, Jan.15 was a Saturday and the office was closed, unlike the seismic office in Shou's hometown Hangzhou which is open every day although there is no earthquake in its history.

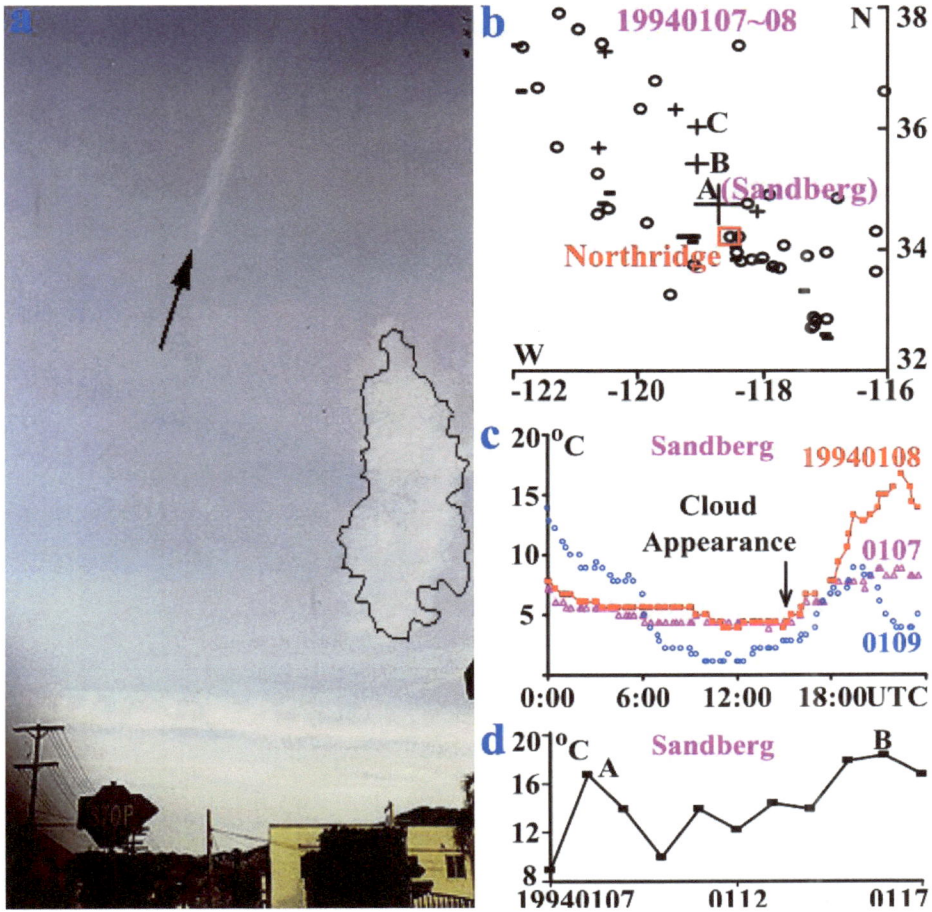

Fig.5 The Northridge earthquake cloud and its accompanying temperature changes on Jan.8, 1994

(**a**) The Northridge earthquake cloud (arrow) appeared suddenly in the northwest sky of Pasadena, California at LT 7:15 am on Jan.8, 1994. Then, the cloud moved quickly northeastward and disappeared at 7:50 am. At 7:30 am, Shou photographed the cloud from the South-East corner of the Green Street - South Chester Avenue intersection in Pasadena toward northwest, using surface buildings as landmarks. A meteorologist from the University of California, Los Angeles, whose field was special clouds, admitted that it did not seem like a weather cloud because it looked like a launching rocket. The black circle indicates an earthquake fog.

(**b**) Daily maximum temperature changes from Jan.7 to Jan.8, 1994 near the Northridge epicenter: The Northridge epicenter is marked with a red hollow square. Plus sign, minus sign, and circle indicate an increase, a decrease, and insignificant change ($<\pm 2^{\circ}C$) in daily maximum temperature, respectively. Pluses marked by A (Sandberg), B (Bakersfield Meadows Field) and C (Porterville) indicate a temperature rise of 7.8°C, 3.9°C, and 3.4°C, respectively. Unmarked plus signs indicate temperature increase of 2~3°C. Small minus signs indicate temperature decrease of 2~3°C; a large minus sign indicates temperature decrease of 3~4°C. The earthquake data are from the USGS@2. All temperature data of **b**~**d** are from the National Climate Data Center (NCDC@3)

(**c**) Temperature profile at Sandberg on Jan.7~ 9, 1994: The Northridge cloud emerged at L.T. 7:15 am, or UTC 15:15 (arrow). on Jan.8, 1994

(**d**) Daily maximum temperature at Sandberg on Jan.7~17, 1994: Temperature peaks A coincided with the Northridge cloud appearance and B one day before the earthquake.

@2 ftp://hazards.cr.usgs.gov/weekly

@3 http://www.ncdc.noaa.gov/oa/ncdc.html

At LT 4:30 am on Jan.17, an M6.7 earthquake hit Northridge, 37 km northwest of Pasadena. The prediction is correct in time, location, and magnitude windows. The quake has been the only one satisfying the predicted location and magnitude windows since Feb.10, 1971. The unique appearance of the cloud, which is unexplainable by meteorology, and the high coincidence between the cloud and the earthquake suggest that the cloud arose from the 18.4 km-deep Northridge hypocenter. Fig.5a also reveals an earthquake fog (circled).

After appearance of the Northridge cloud on Jan.8, 1994, temperatures in areas north to the epicenter experienced an increase (Fig.5b). Sandberg (A), Bakersfield Meadows Field (B), and Porterville (C) had the highest increases of 7.8, 3.9 and 3.4°C respectively in spite of their high altitudes, while other locations had insignificant or small temperature changes. There were no reports of big fires in Southern California at that time. The pattern of temperature increase can be explained by the enormous heat released by the Northridge cloud. The cloud moved northeastward. Locations A, B and C were in the downwind direction along the Sierra Nevada range, and had a distance of 62, 144 and 208km from Northridge respectively. Consisting with movement of the cloud, the location downwind of and closest to the Northridge epicenter experienced the highest increase in daily maximum temperature.

Air is highly insulating, and therefore daily maximum temperature at locations in the upwind direction near the nozzle can remain nearly constant. Further, as hot vapor rising up quickly due to its high pressure, the cold air above descends in a convection which can reduce temperature in neighboring regions not directly downwind of the epicenter. The highest temperature increase, estimated to be 300~1520°C at the nozzle (1.3 High temperature), was not recorded. This is presumably because (1) no station was situated at or very close to the nozzle; (2) hot vapor always rises up quickly due to its high pressure; (3) air is an excellent insulator; and (4) temperature may be too abnormal to be recorded. For example, Sandberg regularly records temperature at :00, :08 and :28, but no records for UTC 17:08, 17:28, and 22:08 on Jan.8, after the appearance of cloud at 15:15 (Fig.5c).

The kinetics of temperature change in Sandberg (Fig.5b, plus A), the station downwind of and closest to the Northridge epicenter, provide further evidence for the enormous heat released by the Northridge cloud. The Northridge cloud emerged at UTC 15:15 or LT 7:15 on Jan.8, and afterward, the temperature rose from 5°C to 17°C (Fig.5c, red solid squares). In contrast, the temperatures did not change nearly as much on Jan.7 or Jan.9 (Fig.5c).

Furthermore, the daily maximum temperature at Sandberg rose from 8.9°C on Jan.7 to a peak of 16.7°C on Jan.8 (A in Fig.5d), the date of cloud appearance. Afterward, daily maximum temperature decreased before it increased to another peak (B in Fig.5d) on Jan.16, one day before the Northridge earthquake. The high temporal coincidence between appearance of the Northridge cloud and temperature increase indicates the enormous heat from the earthquake. The fact that the second temperature peak B so closely preceded earthquake means that it is possible to narrow the time window into a week

1.7.2 Bam Earthquake

At UTC 2:00 on Dec.20, 2003, a cloud suddenly appeared from a fixed nozzle at Bam, Iran (Fig.6, arrow). The cloud was continuously generated from the nozzle while simultaneously. moving eastward for 26 hours. Meteorology cannot explain this phenomenon. Using this cloud, Shou predicted to the public, at UTC 0:58 on Dec.25. an earthquake of magnitude 5.5 or above within 60 days at the vapor source between A and B (Fig.7a, Appendix1) @5.

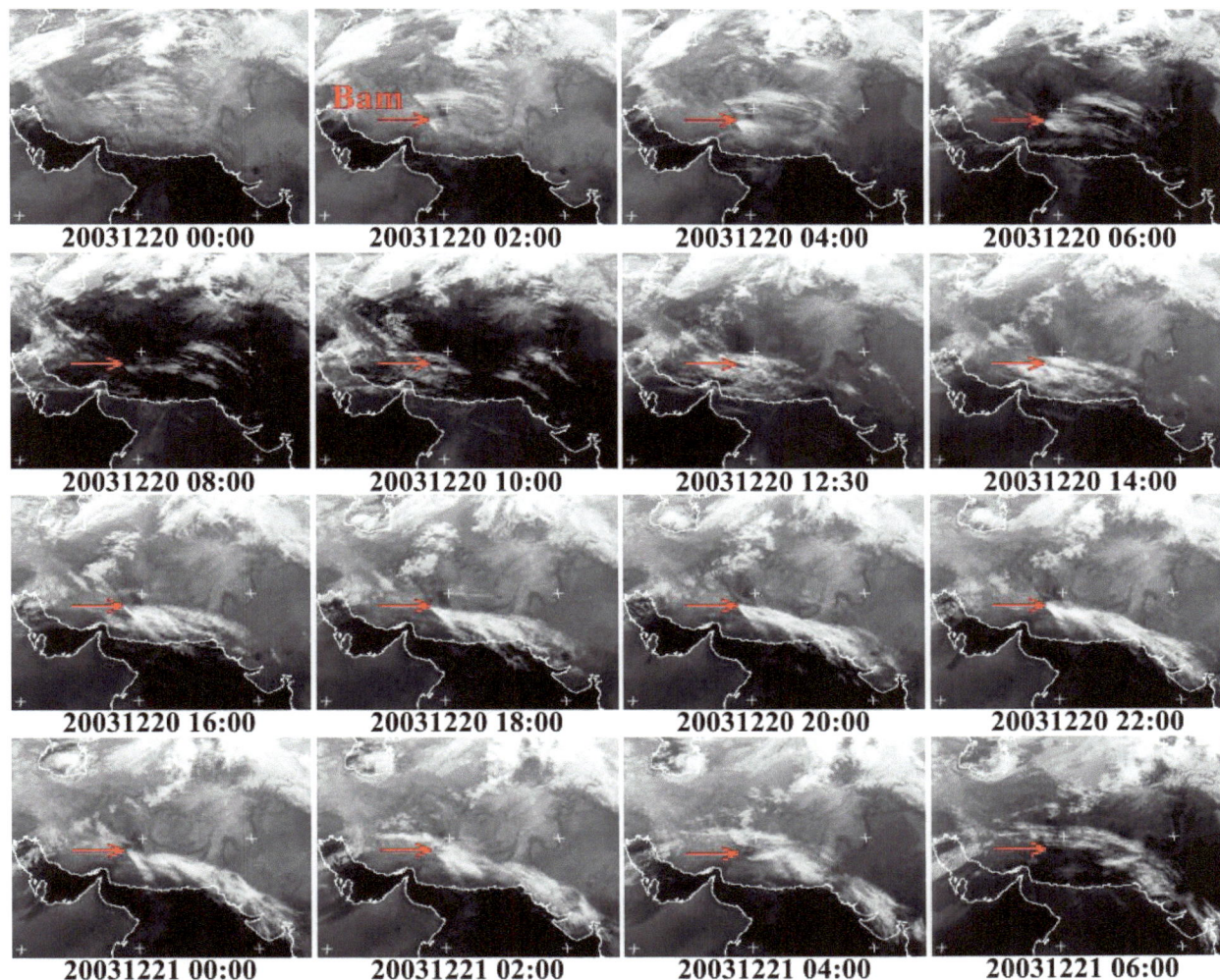

Fig.6: The Bam earthquake cloud on Dec.20~21, 2003
This series of infrared images were initially obtained from Dundee University (DU@4). Neighboring crosses are separated by 10° in latitude or longitude. Arrow indicates vapor nozzle. Coordinates can be found in Fig.7a.
@4 http://www.sat.dundee.ac.uk/pdus.html

The next day, an M6.8 earthquake at Bam proclaimed the success of Shou's prediction (Harrington and Shou, 2005). This earthquake was the only one in the predicted area and magnitude windows in the historical record of about 3,000 years. An animation of the Bam cloud can be found on @6.

@6https://www.youtube.com/watch?v=vC-qmbONlxY

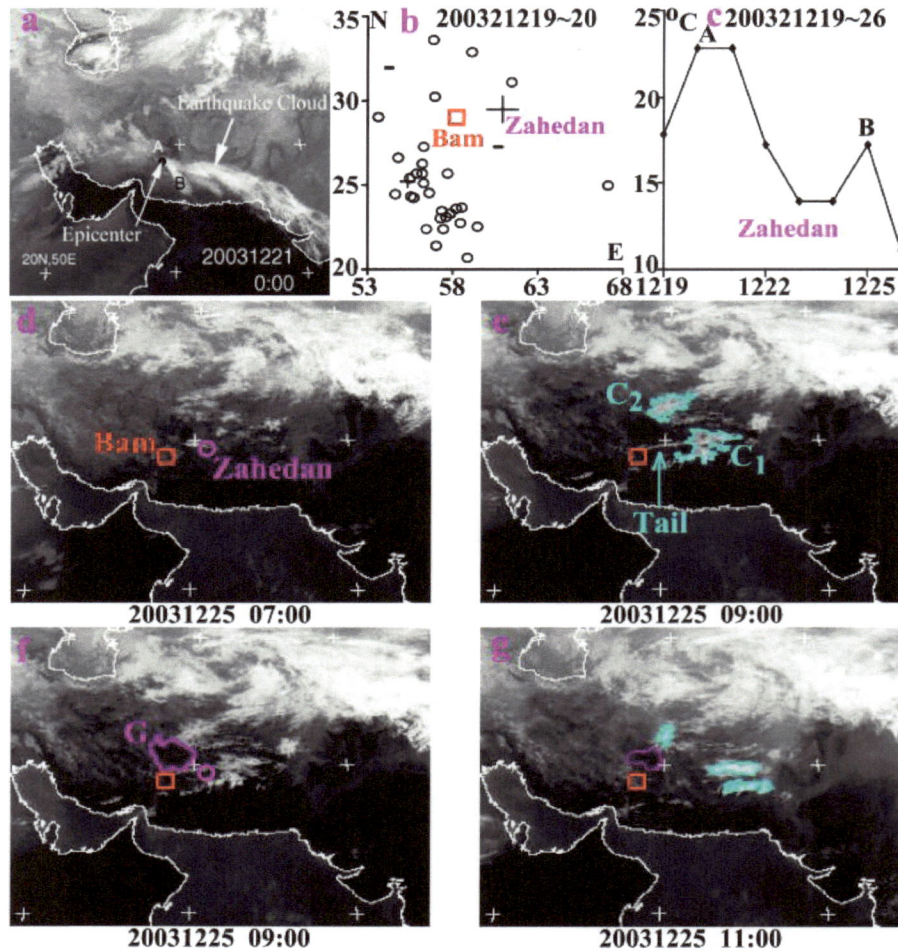

Fig.7. Shou's accurate Bam prediction, temperature changes around and after the Bam cloud, and the second Bam cloud

(**a**) At UTC 0:58 on Dec. 25, 2003, Shou predicted an earthquake of magnitude 5.5 or greater within 60 days between points A and B in the above image. The prediction was announced to the public on his website "Earthquake Clouds & Short Term Predictions" (previous web address: @5a; or present site: @5b; Appendix1). At UTC 1:56 on Dec.26, an M6.8 earthquake hit point A (28.99, 58.29) exactly. This image is from Harrington and Shou (2005) .

(**b**) Daily maximum temperature from Dec.19 to 20 increased in Zahedan, a downwind station, after the Bam cloud. The large plus sign indicates an increase of 5°C at Zahedan. The small plus signs indicate temperature increases of 2~3°C. Circles indicate insignificant temperature changes (<± 2°C). Minus signs indicate temperature changes of -3 ~ -2°C.

(**c**) Daily maximum temperature at Zahedan on Dec.19~26, 2003: Peak A coincided with the Bam cloud date. Peak B was one day before the Bam earthquake.

(**d~g**) The second Bam earthquake clouds: Earthquake cloud C_1 and C_2 (cyan) and Geoeruption G (magenta) appeared around Bam (red square) and Zahedan (magenta circle) at 9:00 on Dec.25, 2003. This temporally coincided with the second temperature peak B in Fig.7c. The thin tail of C_1 (cyan arrow) coincided at the Bam epicenter. The initial images, earthquake data and temperature data are respectively from DU, the USGS and the NCDC.

@5a http://quake.exit.com @5b http://eqclouds.wixsite.com/predictions, http://www.earthquakesignals.com/

During the eruption of Bam earthquake cloud, the temperature at Zahedan, a station downwind of the Bam cloud, increased by 5°C (Fig.7b). It was the highest increase around Bam. The daily maximum temperature at Zahedan rose from 17.8°C on Dec.19, 2003 to a peak of 22.8°C on Dec.20 (A in Fig.7c) during the appearance of the Bam cloud. Then, daily maximum temperature decreased before increased to another peak B on Dec.25 (Fig.7c), one day before the Bam earthquake. This characteristic looks very similar to that of Fig.5d.

After the initial eruption, some water and vapor may remain in the impending hypocenter. This remaining water and vapor can vaporize and erupt again to form the second round of earthquake clouds, geoeruptions, or both. Simultaneously, a temperature peak may result. For example, temperature peak B in Fig.7c occurred on Dec.25, 2003 when earthquake clouds C_1 and C_2 (Fig.7e, cyan) appeared near Bam and Zahedan at UTC 9:00. Note that the tail of earthquake cloud C_1 pointed directly at the impending epicenter Bam. In addition, part of an existing cloud was cleared (Fig.7f magenta, marked as G). Harrington and Shou (2005) hypothesized that this clearing was caused by the heat from earthquake vapor and denoted it geothermal eruption or geoeruption. Geoeruption is an additional hallmark that distinguishes an earthquake cloud from a weather cloud, even though not all earthquake clouds are associated with geoeruptions.

1.7.3 Kerman Earthquake

A linear cloud, outlined in cyan (C in Fig.8b and 8c), appeared from Kerman at UTC 8:00 on Dec.14, 2004. Its tail pointed toward the impending epicenter (red square). The cloud grew to a long straight line and its middle part started to bend northward at 10:00 under the influence of a south wind. The sudden appearance, quick growth, and linear shape of the cloud suggested it as an earthquake cloud. The length of the cloud was 370km, given the distance of 960km between (30, 60) and (30, 70). For linear earthquake clouds, the magnitude of earthquake is proportional to the length of cloud (Shou, 2006a). This greater-than-300km cloud indicates an earthquake of magnitude greater than 6. The cloud changed its shape due to south wind and disappeared after 16:00.

Simultaneously, Geoeruption G_1 formed in clouds at 10:00 (Fig.8c, Magenta). This geoeruption persisted near Kerman until 16:00, and predicted the epicenter nearby (red square). The M6.5 Kerman earthquake was the only earthquake of magnitude 6.5 or above in the area of 20~50N and 40~70E from Dec.27, 2003 to Oct.27, 2008.

Additional moderate geoeruptions and earthquake clouds could be found in Fig.8. For example, emerging in cloud at 8:00, G_2 (outlined in magenta) was also a geoeruption. By contrast, large black places marked by "N" were not in clouds, and were therefore not geoeruptions. South wind blew the Kerman earthquake cloud northward.

In downwind direction, the average temperature of an area marked by points MGABSTGy (44480km^2 in Fig.8g) increased 4.1°C from Dec.13 to 14, 2004. Ghuchan (G) and Mashhad (M) increased 6.1 and 5.6°C respectively. When the Kerman earthquake cloud appeared, the daily maximum temperature of Mashhad increased from 7.2°C on Dec.13 to 12.8°C on Dec.14 (Peak

A in Fig.8h). Afterward, it decreased and experienced multiple peaks (caused by eruption from other locations, see below) before reaching peak B (Fig.8h), two days before the earthquake.

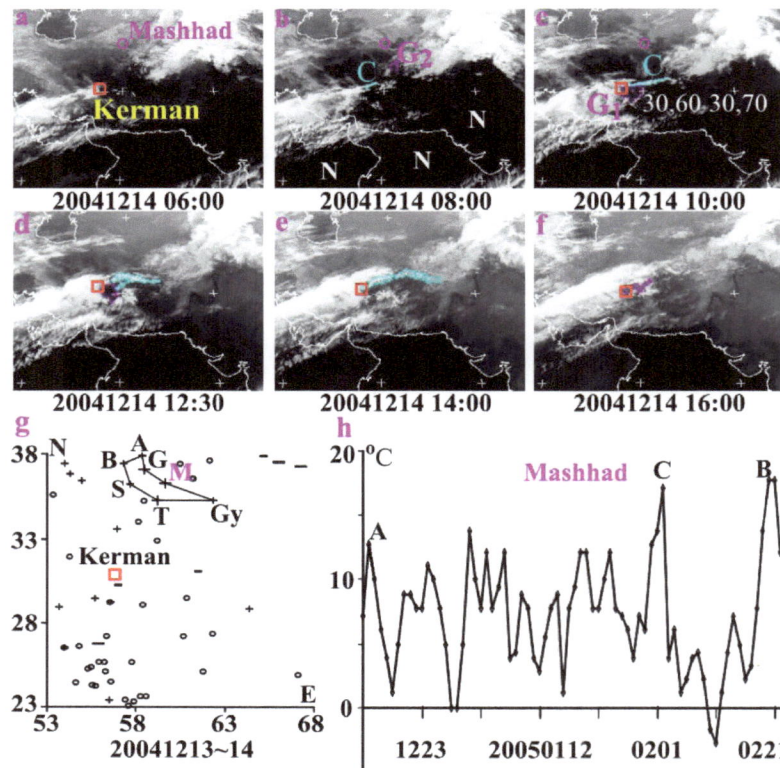

Fig.8 The Kerman earthquake cloud on Dec.14, 2004 and accompanying temperature change
(**a~f**) Red square points at the M6.5 Kerman epicenter (30.74, 56.83) on Feb.22, 2005. Magenta circle plots the Mashhad station (36.26, 59.63). Cyan outlines an earthquake cloud C, while magenta outlines geoeruptions G_1 and G_2. "30, 60" and "30, 70" are "30N, 60E" and "30N, 70E," respectively. Black regions marked with "N"s are not geoeruptions.
(**g**) Daily maximum temperature change from Dec.13 to 14, 2004 in areas near Kerman: Red square plots the Kerman epicenter. "o" represents a temperature change of $\le \pm$ 2°C. Small, medium, and large "-" or "+" represent a temperature reduction or increase of 2.0~3.9°C; 4~5.9°C, and >6°C respectively. A (Ashgabat Keshi), B (Bojnourd), G (Ghuchan), M (Mashhad), S (Sabzevar), T (Torbat-Heydarieh), and Gy (Gyshgy) are stations downwind of the Kerman earthquake cloud. They enclose an area of 44480km^2 where average temperature increases 4.1°C from Dec.13 to 14, 2004.
(**h**) Daily maximum temperature of Mashhad from Dec.13, 2004 to Feb.22, 2005: Peaks A, B, and C occurred on Dec.14, 2004, Feb.21, 2005, and Feb.2, 2005 respectively. The initial images, earthquake data and temperature data are from DU, the USGS and the NCDC respectively.

Like the second Bam eruption (Fig.7e~g) which was temporally correlated with the second temperature peak at Zahedan (Fig.7c), the second Kerman eruption (Fig.9) is temporally correlated with the temperature peak B at Mashhad on Feb.20~21 (Fig.8h), a few days before the Kerman earthquake. These large temperature increases (a few days before a large earthquake) look like those in Fig.5d and Fig.7c.

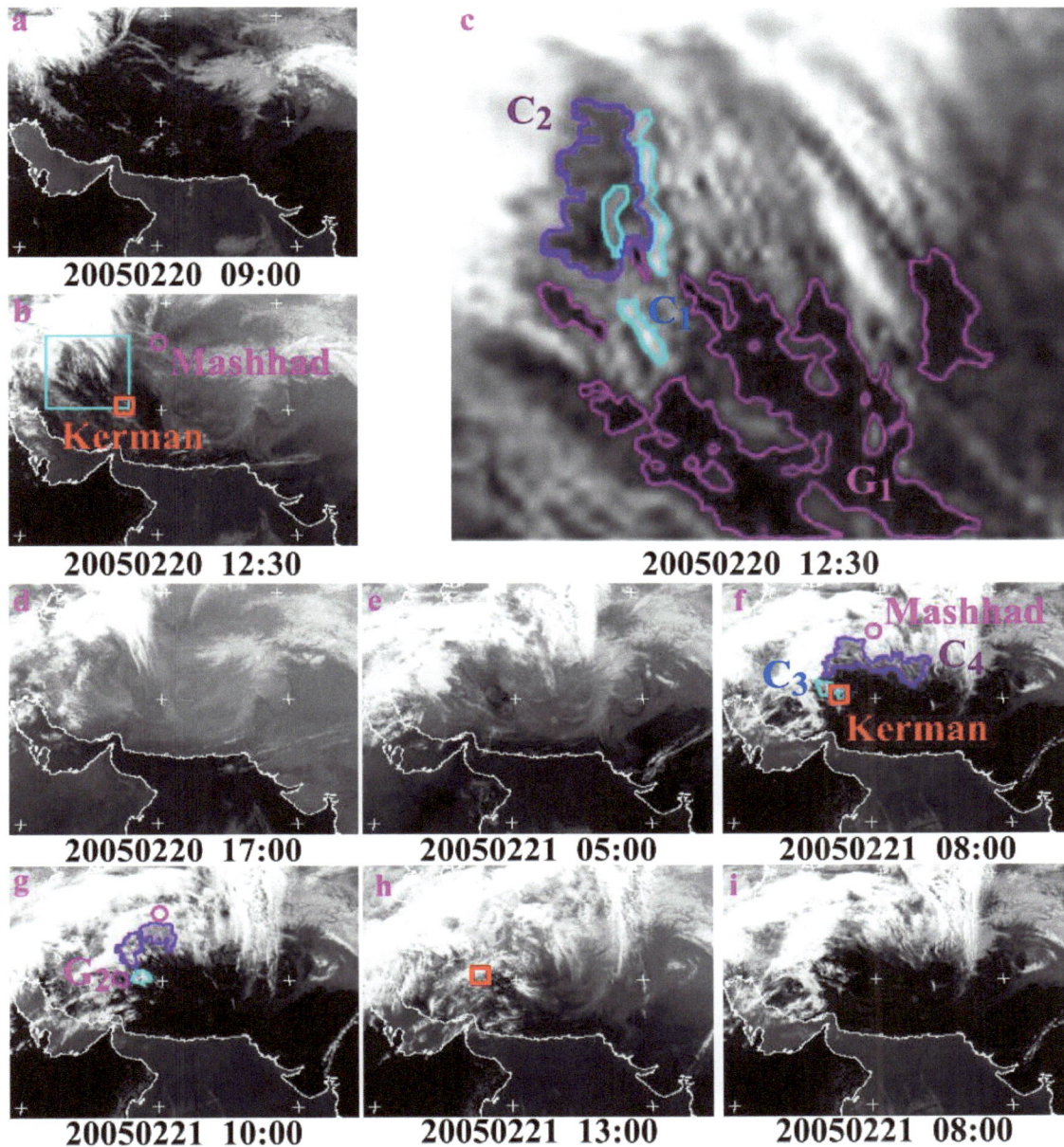

Fig.9: Vapor eruptions in Kerman on Feb.20~21, 2005 vs. temperature peak B in Mashhad
Red square and magenta circle respectively point at the M6.5 Kerman epicenter and the Mashhad station where daily maximum temperature in Fig.8h reach peak B on Feb. 20-21, 2005. The cyan square in panel **b** reveals a vapor eruption near Kerman just on Feb. 20, 2005. Panel **c,** magnified cyan square, shows geoeruption G_1 (magenta edge), part of wave-shaped clouds C_1 (cyan edge) and part of loose clouds C_2 (purple edge). Panels **a** and **d** offer images before and after panel **b** respectively. Panels **f** and **g** depict earthquake clouds C_3 (cyan), C_4 (purple) and georuption G_2 (magenta). Panels **e** and **h** are images before and after panels **f** and **g** respectively. Panels **i** is an unmodified image of panel **f.** All initial images are from DU.

Fig.9 reveals two eruptions: one on Feb.20, 2005 and the other on Feb.21. They are temporally correlated with the temperature peak B at Mashhad (Fig.8h). The former (Fig.9c) contains geoeruption G_1 (magenta edge), wave-shaped clouds C_1 (cyan) and loose clouds C_2 (purple).

Wave-shaped clouds can be explained by vapor through a row of nozzles. Chapter 2 will discuss them further. Loose clouds could form when earthquake fog condenses, or when vapor heat incompletely dissipates preexisting clouds, or both. Fig.9c only outlines part of the both clouds. Readers can find more. The latter (Fig.9e~i) also contains wave-shaped cloud (C_3), loose cloud (C_4) and geoeruption (G_2) near the epicenter

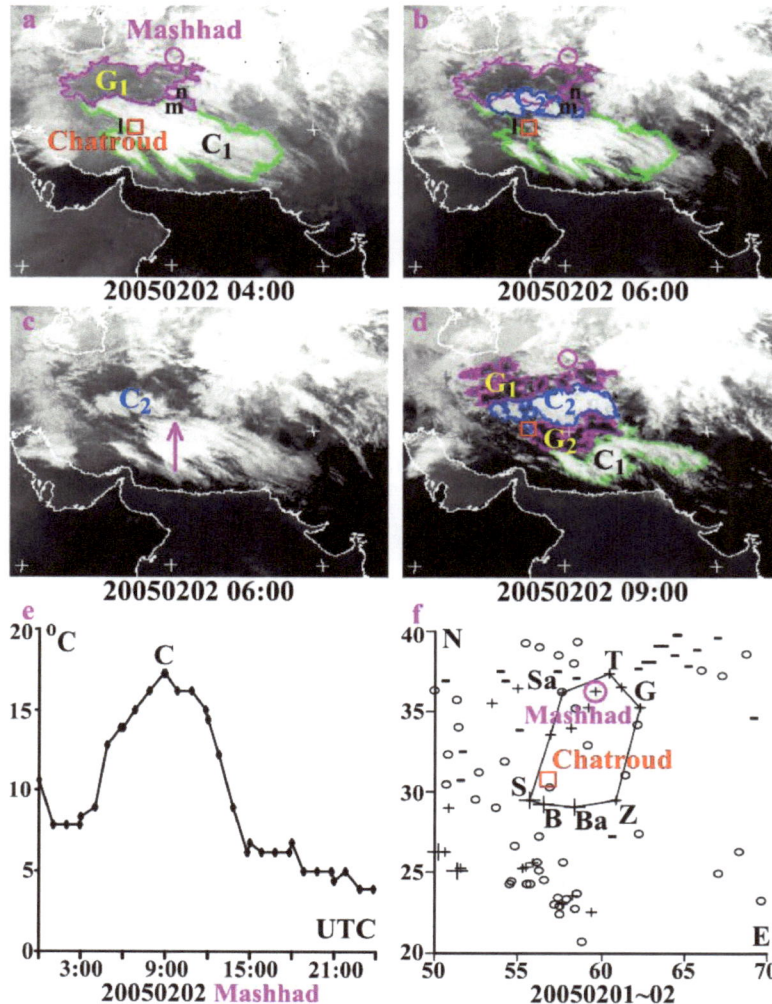

Fig.10: Vapor eruption in Chatroud vs. temperature peak C at Mashhad on Feb.2, 2005
Red square and magenta circle respectively point at the M5.5 Chatroud epicenter (30.58, 56.86) on May 14, 2005 and the Mashhad station where daily maximum temperature on Feb. 2, 2005 reaches peak C in Fig.8h. Magenta and green in Panel **a** outline geoeruption G_1 and cloud C_1 respectively. The same outlines were copied to Panel **b** which also included a blue outline for earthquake cloud C_2. Panel **c** is identical to Panel **b**, except without any outlines. The arrow in Panel **c** points at a black line that separated cloud C_2 from C_1, and developed into a geoeruption G_2 (Panel **d**). Part of clouds marked by l, m and n at 4:00 (Panel **a**) disappeared by 6:00 (Panel **b**). Panel **e** reveals detail of temperature increase of Mashhad on Feb.2, 2005. It increases 8.3°C from UTC 4:00 to 9:00 that coincides the eruption time from UTC 4:00 to 9:00 (Panels **a~d**). Panel **f** plots stations S (Sirjan), B (Baft), Ba (Bam), Z (Zahedan) G (Gyshgy), T (Tedzhen) and Sa (Sabzeva). They enclose an area of 338,000km² where average temperature increased 3.6°C from Feb.1 to 2, 2005. Temperature data are from the NCDC. Initial images are from DU

Among the many temperature peaks between peaks A and B, the highest peak C (Fig.8h) occurred at 9:00 on Feb.2, 2005 (Fig.10e). However, peak C is caused by the vapor eruption preceding the M5.5 Chatroud earthquake (30.58, 56.86) on May 14, 18km away from the epicenter of the M6.5 Kaman earthquake (30.74, 56.83) on Feb.22.

Specifically, on Feb.2, temperature of the Mashhead station started to rise at 4:00 (Fig.10e). At about the same time, geoeruption G_1 appeared (Fig.10a). Then cloud C_2 appeared (Fig.10b; an unmarked version of the same image in Fig.10c to show the natural outline of this cloud). This cloud moved northward against a wind that had pushed another cloud (C_1, green outline) southeastward. Cloud C_2 erupted in a direction independent of the wind direction, suggesting that C_2 was generated under high pressure. This is an important characteristic that can distinguish an earthquake cloud from weather clouds. By this evidence, C_2 is an earthquake cloud, and its source should locate at somewhere south of C_2.

Images at 4:00 and 6:00 have the same magenta and green outlines. Thus, it is easy to find parts of clouds at 4:00 that had disappeared by 6:00 (e.g. locations marked with l, m, and n). These dissipated cloud parts were melted by the heat of cloud C_2. The arrow in panel **c** pointed to the boundary separating C_2 from C_1. This boundary was also formed by C_2, and became a large geoeruption G_2 at 9:00 (panel **d**). This phenomenon of geoeruption is yet another important characteristic to distinguish cloud C_2 from weather clouds. C_2 became bigger until 9:00. Afterward, it became smaller and finally disappeared. Thus, temporally and spatially, the appearance of earthquake cloud C_2 and its associated geoeruptions coincided with the temperature increase in Mashhead station which is downstream of the direction of eruption.

Earthquake cloud C_2 released an enormous amount of heat which warmed not only Mashhead, but also a vast space between Mashhead and the eruption center. This space included an area of 338,000km^2. In this area, the average temperature increased 3.6°C from Feb.1 to 2, 2005 (Fig.10f).

The size of the earthquake cloud suggested a magnitude 5.3~5.7 earthquake. Within the longest duration of 112 days from Feb.2 to May 25, 2005, only the M5.5 Chatroud earthquake near Kaman on May 14 was consistent with the cloud. In this case, the Chatroud earthquake occurred more than three months after the appearance of the earthquake cloud and geoeruption. Thus, even though between temperature peaks A and B, which are respectively caused by the primary and secondary Kerman earthquake eruptions, many other temperature peaks existed (Fig.8h), these peaks are likely to be caused by eruptions from epicenters not identical to the M6.5 Kerman epicenter.

Monitoring temperatures at a fixed epicenter can more accurately isolate temperature increases due to particular eruptions from that epicenter, which can in turn be used to narrow down the time window of prediction since an earthquake usually occurs a few days after the second vapor eruption and temperature peak.

1.7.4 Turkey Moderate Earthquakes

Fig.11 presents an example of geoeruptions associated with a swarm of moderate earthquakes in Turkey. In Panel **a**, a geoeruption emerged from Point A at 8:00 on Feb.23, 2000 and disappeared by 15:00. Another geoeruption appeared from X at 15:00 and erupted westward. The tail of geoeruption gradually extended northeastward to form a black band XBC. Along this black band, two bulges formed: one at B (21:00), and the other at C (22:00). A magnified view of both bulges was presented in Panel **g**. The geoeruption disappeared by 2:00 the next day. In Panel **h**, a schematic hypothesizing the formation of X, B, and C was presented. Point X was a nozzle, the exit of a main crevice. Eruptions from hypocenters B and C mainly escaped from X. through the main crevice. Some eruption escaped through finer crevices (black C_1, green C_2 and blue C_3 dotted lines), forming a black band XBC, and Bulges B and C

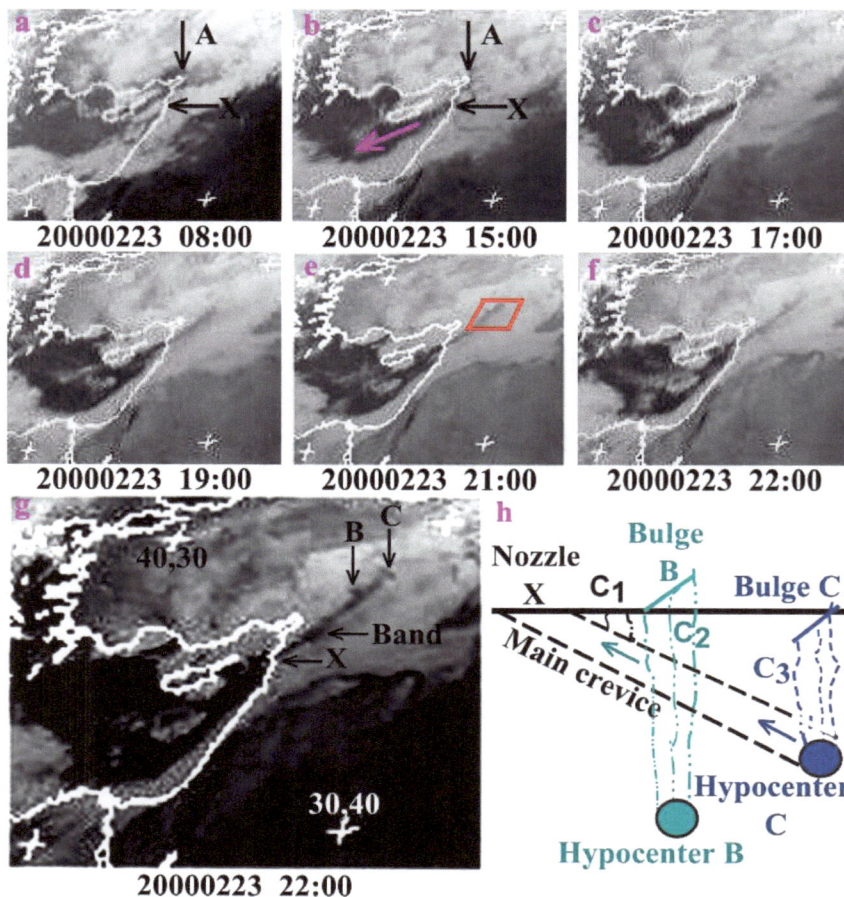

Fig.11: Turkey geoeruptions on Feb.23, 2000 and their schematic diagram.
(**a~f**) A geoeruption occurred at A (37.1~37.3, 35.1~37) at UTC 8:00 on Feb.23, 2000 and disappeared by 15:00. Another warm spot appeared at X (~35.2, 35.9), causing a geoeruption southwestward (magenta arrow). Meanwhile, the tail of this geoeruption grew northeastward. At 22:00 in the tail region, two small bulges formed at B (37.6~37.9, 37.2~37.3) and C (38.1~38.4, 38.3~38.8). The red rectangle was the coarse area window in Shou's prediction to the USGS on Feb.28, 2000 (Harrington and Shou, 2005).
(**g**) The magnified image at 22:00 reveals more details of X, bulges B and C and a band connecting them. The original images were from DU.
(**h**) A schematic diagram of the formation of Nozzle X, and bulges B and C. Nozzle X is the exit of the main crevice. Small crevices C_1, C_2 and C_3 respectively connected the main crevice, and hypocenters B and C to the surface. Erupting vapors from hypocenters B and C mainly passed through the main crevice and exited through nozzle X, so eruption (black indicating high temperature) appeared at X at first. Simultaneously, part of the vapor passed through small crevices C_1, C_2 and C_3 to the surface to form the black band XBC. Because C_1 was shorten than C_2 and C_3, the tail of geoeruption grew from X to B and C. C_2 and C_3 contained many crevices, and vapors escaped through these crevices formed bulges B and C respectively.

Based on the above hypothesis, Shou predicted to the USGS on Feb.28, 2000 that there would be a magnitude 5 or two magnitude 4 earthquakes within a coarse area window of latitude 36.5~38.5, longitude 36~ 39 (rectangle in Fig.11e) and within the 50 days from Feb.28 through Apr.18. In the same prediction, a finer area window of latitude 37~37.8, longitude 36.8~37.2 (too small to show) and a finer time window of 17 days from Mar.25 to Apr.10 were also made. The prediction was correct for the finer time and area windows, as two earthquakes of magnitudes 4.3 and 4.5 occurred on Apr.2 coinciding with bulge B, well within the coarse and at the edge of the fine area window. Except this pair, neither a magnitude 5, nor two magnitude 4 earthquakes have occurred in an area (37~38, 36.8~38) containing the fine area window for more than 22 years from the beginning of the database on Jan.1, 1990 to Sep.18, 2012. Within the fine time window, the predicted pair were the only earthquakes bigger than or equal to 4 in a region of (29~44, 31~48), a region 637 times larger than the predicted fine area window (Harrington and Shou, 2005).

Table 2: Earthquakes associated with the Turkey geoeruptions in Fig 11.

Geoeruptions					Earthquakes					
Date UTC	Time	P	Lat. N	Lon. E	Date UTC	Time	Lat. N	Lon. E	Mag. M	Dep. Km
20000223	8:00	A	37.2	36.0	000512	3:01	37.04	36.08	4.8	10.0
	21:00	B	37.7	37.3	000402	11:41	37.63	37.32	4.5	9.0
					000402	17:26	37.62	37.38	4.3	9.0
	22:00	C	38.3	38.5	000507	9:08	38.18	38.74	4.4	1.6
					000507	23:10	38.16	38.77	4.5	5.4

Note: P: Point in Fig.11; Lat: Latitude; Lon: Longitude; Mag: Magnitude: Dep: Depth. The average latitudes and longitudes of geoeruptions A, B and C are shown in the table. The differences between the latitude and longitude of an earthquake and the average latitude and longitude of the corresponding geoeruption are 0.1° and 0.1°, respectively. Earthquake data are from the USGS.

Earthquakes also occurred later at points A and C, again coinciding with geoeruptions (Table 2). In a 2°-long square centered at A, the M4.8 earthquake on May 12 was the only quake of magnitude 4.5 or above within 586 days from Jun.11, 1999 to Jan.16, 2001. In a 2°-long square centered at C, the two M4 earthquakes on May 7 were the only two magnitude 4 or above earthquakes within 2750 days from Feb.8, 1996 to Aug.19, 2003. The high spatial coincidence between geoeruptions and their associated earthquakes suggests geoeruptions being formed by erupting vapors from impending hypocenters.

Neither geoeruption C nor nozzle X had nearby downwind stations. In contrast, stations Adana/Vincirlik (Ad) and Kahramanmaras (Ka) were downwind from and near geoeruption A and B respectively (Fig.12a). From Feb.22 to 23, 2000, temperature of Ad and Ka increased 3.9°C and 2.3°C respectively; temperature of Basel Assad, a station near but not downwind from the Nozzle X increased slightly by 0.5°C. In contrast, the average temperature of the area spanning 30~40N and 30~40E decreased 2.4°C. This contrast suggests earthquake vapor containing enormous heat. Like the Northridge earthquake, the Bam earthquake, and the

Kerman earthquake, the M4.8 Turkey earthquake on May 12 and the two M4 Turkey earthquakes on Apr.2 occurred within a few days after their respective downwind stations Adana/Vincirlik (Ad) and Kahramanmaras (Ka) had reached a peak temperature (Fig.12b)

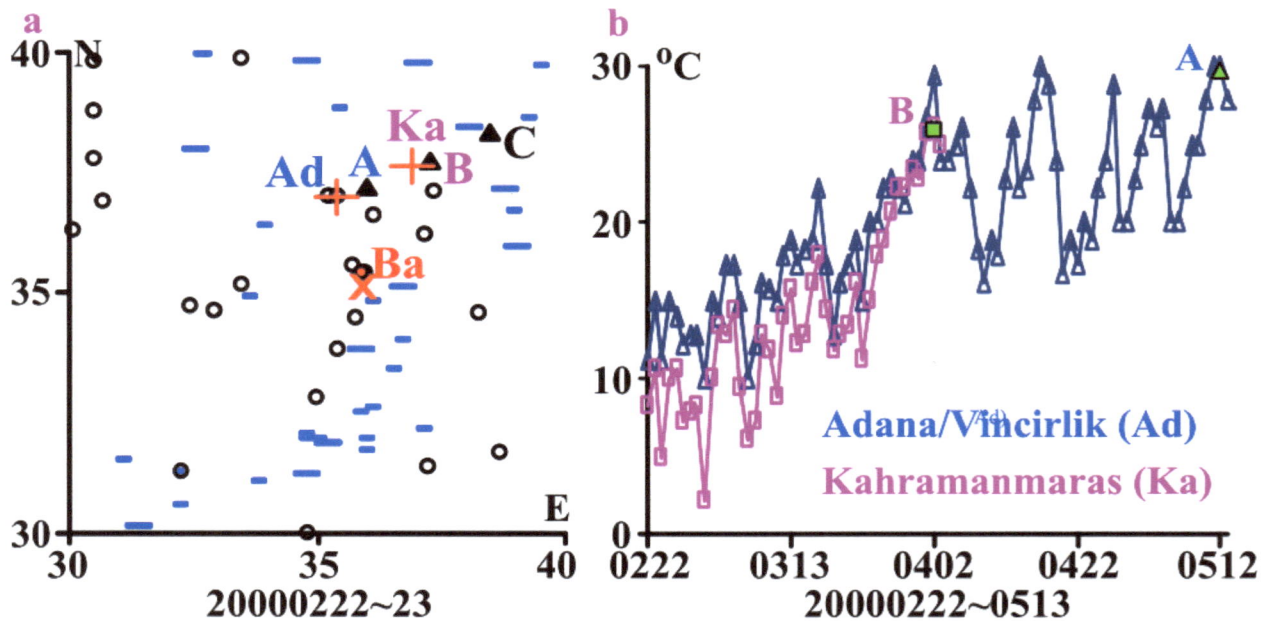

Fig.12: Temperature changes associated with geoeruptions in Turkey from Feb.22 to Feb.23, 2000.
(**a**) Solid triangles A, B, and C respectively plot geoeruptions A, B, and C in Fig.11 and Table 2. Ad and Ka indicate stations Adana/Vincirlik and Kahramanmaras where a temperature increase of 2.0~3.9°C from Feb.22 to Feb.23, 2000 ("+") was recorded. Small and large "-" represent a temperature reduction of 2.0~3.9°C and ≥4°C respectively; "o" represents temperature changes of <±2°C. Red "X" plots the nozzle X in Fig.11g. Station Ba (Basel Assad, red-center) showed a temperature increase of 0.5°C.
(**b**) Blue triangles and magenta squares respectively show daily maximum temperatures of Adana/Vincirlik (Ad) and Kahramanmaras (Ka) from Feb.22 (one day before geoeruption) to May 13, and Apr.3, 2000. The green-filled square reveals a temperature peak at Ka on Apr.2 when two M4 earthquakes struck nearby. The green-filled triangle reveals a temperature peak at Ad on May 12 when an M4.8 earthquake struck nearby. These characters are similar to those of Figs.5d, 7c, and 8h. The temperature data are from the NCDC.

1.7.5 Indian Tsunami Earthquakes

On Dec.26, 2004, a swarm of eleven big earthquakes hit the East Indian Ocean. Before these earthquakes, vapors erupted and mixed, making it difficult to distinguish vapors from each other. However, a bigger earthquake releases more vapor than smaller earthquakes, and thus, it is possible to find the largest impending epicenters. Among this swarm of earthquakes, three largest ones were the M9.0 Sumatra earthquake (3.30, 95.98, red square C in Fig.13d), the M7.5 Nicobar earthquake (6.91, 92.96, green square B in Fig.13d), and the M6.6 Andaman earthquake (8.88, 92.38, cyan square A in Fig.13d). Those earthquakes erupted vapors on Nov.14, 2004 and formed three long straight lines of clouds (Fig.13e~k). One terminus of each of the three lines coincided with one of the three largest future epicenters. The vapors warmed a large area. Consequently, surface temperatures surpassed 100°C in Karachi and Lahore, Pakistan.

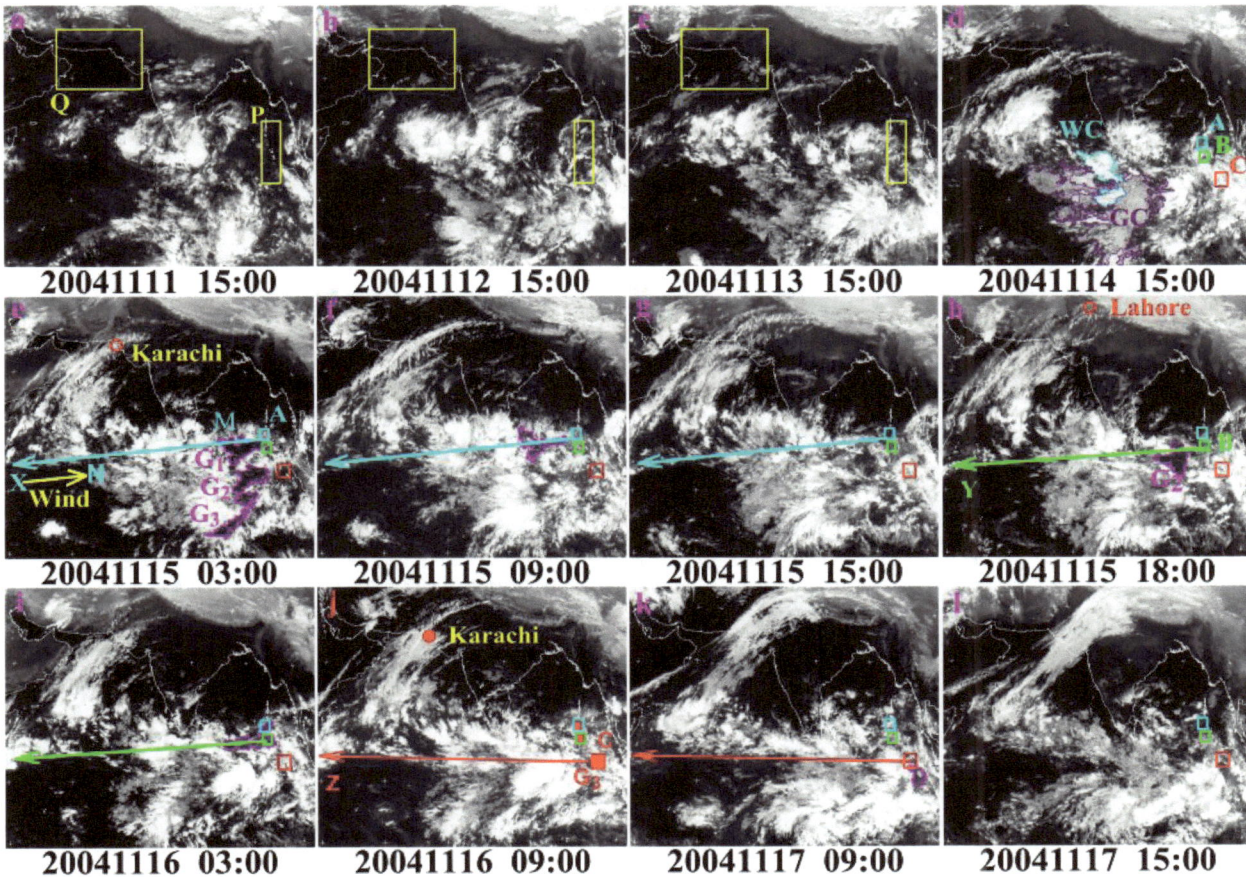

Fig.13: A swarm of Indian Ocean vapor eruptions on Nov.14~17, 2004

(**a**) Rectangle P on the East Coast and rectangle Q on the North Coast of the Indian Ocean were clear of clouds at 15:00 on Nov.11.

(**b~c**) Clouds increased on the East Coast during Nov.12~13.

(**d**) Gray loose clouds GC (purple edge) and white clouds WC (cyan edge) appeared quickly and moved westward at 15:00 on Nov.14 (see animation@7). The both were earthquake clouds. Their large volume suggests a swarm of large impending earthquakes erupting vapors; their moving direction suggests that the source was on the east coast. Squares A (cyan), B (green) and C (red) respectively plot the epicenters of three largest earthquakes all of which struck on Dec.26, 2004: the M6.6 Andaman earthquake (8.88, 92.38), the M7.5 Nicobar earthquake (6.91, 92.96) and the M9.0 Sumatra earthquake (3.30, 95.98).

(**e**) Cyan arrow AMNX depicts a long straight line. Its terminus A was an impending epicenter; AM bordered geoeruption G_1 (magenta triangle); and NX separated clouds and cloudless space. This separation of clouds and cloudless space was due to wind (yellow arrow) that removed weak vapor to form cloudless space. However, the wind was not strong enough to counter the strongest vapor which continued to move westward to form clouds. In Karachi (red circle), temperature reached 225°C at 4:30 (LT 9:30am) on Nov.15@8@9.

(**f-g**) G_1 largely disappeared by 15:00, Nov.15.

(**h-i**) Green arrow BY reveals another long straight line at 18:00. Its terminus B was an epicenter, its east part bordered geoeruption G_2 (magenta triangle), and its west part separated clouds and cloudless space. In Lahore (red circle), temperature reached 146°C at 16:30 (LT 9:30pm) on Nov.15. G_2 became smaller by 3:00 on Nov.16@8@10.

(**j-l**) Red arrow CZ shows the third long straight line appearing at 9:00 on Nov.16. Its terminus C was an epicenter, its east part bordered geoeruption G_3 (magenta point), and its west part separated clouds and cloudless space. In Karachi (red circle), temperature reached 288°C at 13:00 (LT 18:00) on Nov.16@8@11. CZ lasted until about 0:00

- 19 -

the next day, disappearing gradually. The images, earthquake data, and temperature data were from DU, the USGS, and the Weather Underground website (WU@8) respectively.

@7 https://docs.google.com/file/d/0B3PS6mjpf0ITSnN5MHY4NFJncW8/edit?usp=sharing
@8 http://www.wunderground.com/
@9 https://docs.google.com/file/d/0B3PS6mjpf0ITNmhxcUdPNHI2TjQ/edit?usp=sharing
@10 https://docs.google.com/file/d/0B3PS6mjpf0ITNEN6eHllLUFNSG8/edit?usp=sharing
@11 https://docs.google.com/file/d/0B3PS6mjpf0ITTlhCV3oxNnRncmM/edit?usp=sharing

Specifically, areas P and Q of the Indian Ocean were clear of clouds at 15:00 on Nov.11, 2004 (Fig.13a). Clouds increased in area P on Nov.12~13 (Fig.13b~c). At 15:00 on Nov.14, many vapors began to erupt from the east coast to the west to form large gray clouds GC and heavy white clouds WC (Fig.13d). The former was warm and at a low vapor density and a low altitude. The former rose up and condensed into the latter, cold heavy clouds. Their large mass and quick growth suggested a swarm of large earthquakes erupting vapors.

Alternatively, part of the vapors rose up to form geoeruptions. The three largest geoeruptions G_1, G_2 and G_3 were depicted by magenta edges at 3:00 on Nov.15 (Fig.13e). AM of G_1 originated from the epicenter of the later M6.6 Andaman earthquake. A wind heading northeast (yellow arrow, Fig.13e) interrupted weak vapor, so a cloudless space formed south of NX. The wind could not interrupt strong vapor, so clouds increased north of NX, forming a straight boundary between clouds and cloudless space (Fig.13e~g). Coincidently, A, M, N and X were in a long straight line (cyan arrow). This phenomenon suggests that erupting earthquake vapor moved in a straight line and that the M6.6 Andaman earthquake vapor was stronger than the west wind at 3:00 on Nov.15 to form the straight line. The clouds north of NX turned to Pakistan and India, and then disappeared. A red circle plots Karachi (Fig.13e), whose airport recorded a temperature of 225°C at 4:30 (LT 9:30am) on Nov.15@9. Geoeruption G_1 became smaller at 9:00 (Fig.13f) and disappeared by 15:00 (Fig.13g). The original straight line AX persisted.

Similarly, another long straight line BY formed at 18:00 on Nov.15 (green arrow, Fig.13h) and terminus B of geoeruption G_2 coincided with the epicenter of the later M7.5 Nicobar earthquake. A red circle plots Lahore, where temperature reached 146°C at 16:30 (LT 9:30pm) on Nov.15@10. This straight line BY lasted until 3:00 on Nov.16 (Fig.13i). Like AX and BY, the third long straight line CZ formed at 9:00 on Nov.16 (red arrow, Fig.13j) and C coincided with the epicenter of the M9.0 Sumatra earthquake was very close to a small geoeruption G_3. A red circle plots Karachi (Fig.13j) where temperature reached 288°C at 13:00 (LT 6:00pm) on Nov.16@11. The straight line CZ lasted until 0:00 the next day and then disappeared gradually. An animation shows the entire process@7.

The M6.6 Andaman earthquake has been the largest in the area of 8~10N, and 90~95E since Jan.1, 1990, the beginning of the ftp-accessible curated USGS database @2. The M7.5 Nicobar earthquake was the only earthquake of magnitude more than or equal to 7 in the area of 5~20N and 90~105E within 15 years from Jan.1, 1990 to Jul.23, 2005. The 2004 M9.0 Sumatra earthquake was the largest one in the world in the 47 years spanning after 1964 (when the M9.2 Alaska earthquake struck on Mar.27) to 2011 (when the M9 Japan earthquake struck on Mar.11).

The high coincidence between the three long straight lines and the three rare large earthquakes strongly support the earthquake vapor model.

The vapors of the M6.6 Andaman earthquake, the M7.5Nicobar earthquake, and the M9.0 Sumatra earthquake erupted together at 15:00 on Nov.14, and lasted about 24, 36 and 57 hours respectively. Thus, the duration of eruption correlates with magnitude of the earthquake.

Peak surface temperatures reached above $100^{\circ}C$ in Karachi and Lahore, strongly supporting the earthquake vapor model. Because the shortest distance between the three earthquakes and the two Pakistan airports is 3,100km, one can deduce that a large area around the Northern Indian Ocean has abnormally high temperature.

1.8 Definition of Abnormal Temperature

Shou's preliminary definition (Shou, 2011) for "abnormal" is to satisfy at least one of the following:

    (1) Air temperature reaches or surpasses $60^{\circ}C$ (the highest in meteorology);

    (2) The trend in hourly temperatures shows a pulse increase such as point H in Fig.14a.

    (3) The daily maximum temperature reaches the highest in a month, such as point D in Fig.14b;

    (4) The daily maximum temperature of a day reaches or surpasses daily maximums of the same day in many years as denoted by point D_1 in Fig.14c;

    (5) If the highest daily maximum temperature has been proven to result from earthquake vapor as denoted by point D_1 in Fig.14c, then the second highest daily maximum temperature is also considered abnormal (point D_2 in Fig.14c);

    (6) The daily maximum temperature increase is much higher than those of its surrounding area such as the increase of $5^{\circ}C$ at Zahedan (Fig.7b) versus $2.2^{\circ}C$ as the next largest temperature increase in its surrounding.

Abnormal temperatures recorded by Kerman airport

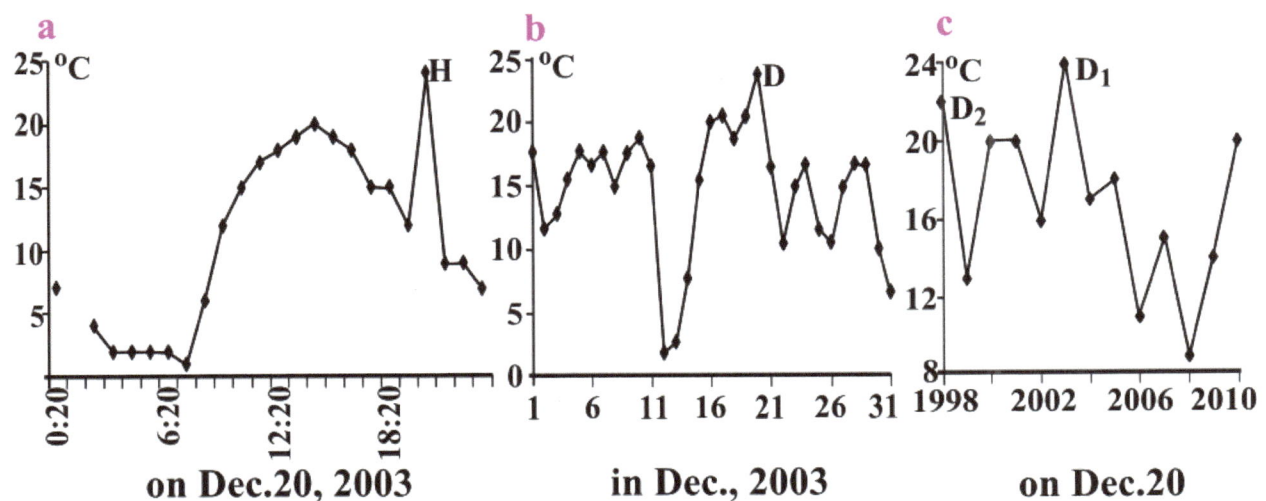

Fig14. Adopted from Fig 5 of US Patent 8068985 (Shou, 2011) Time: LT

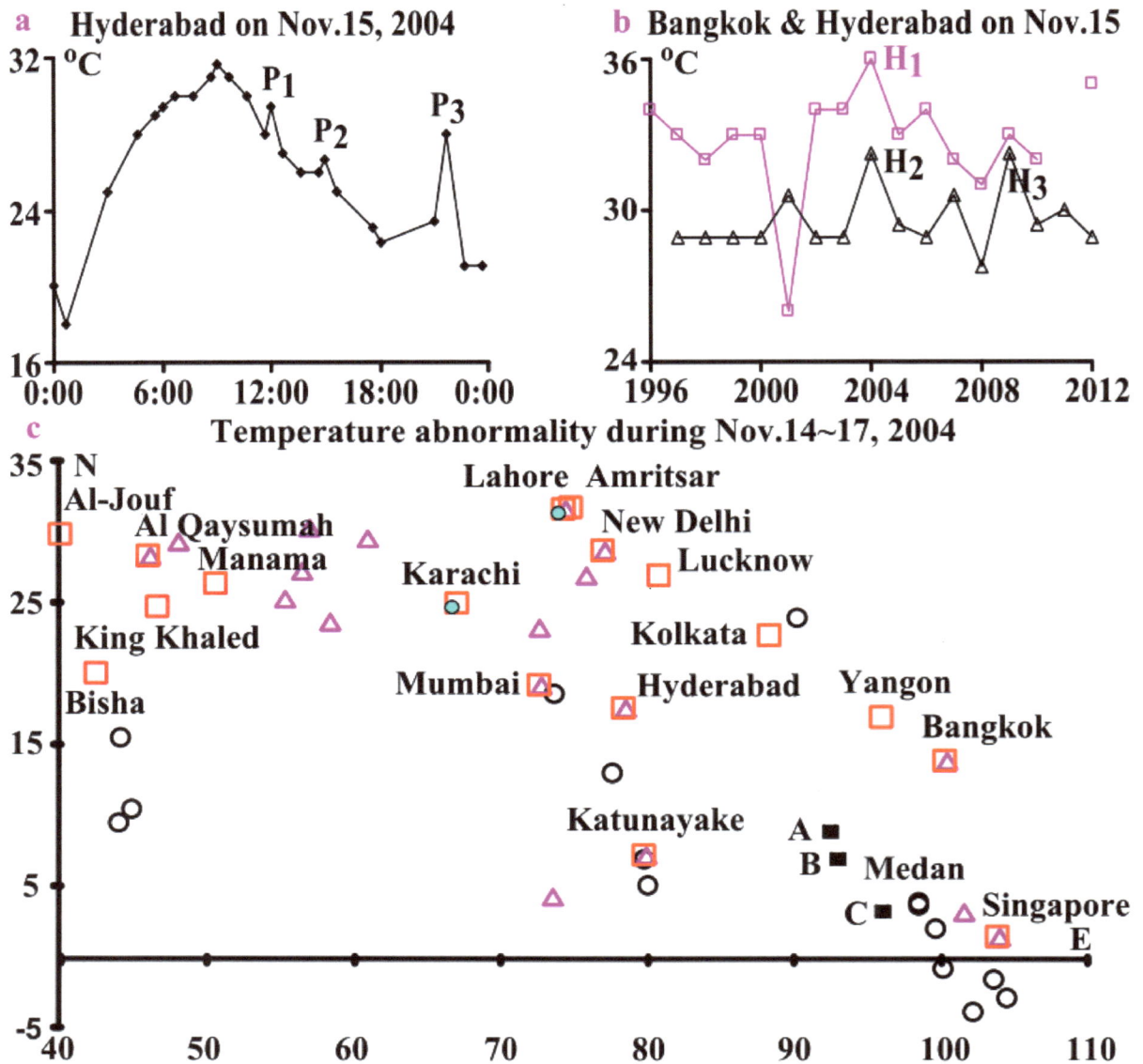

Fig.15: Temperature abnormality during the Indian Ocean vapor eruptions on Nov.14-17, 2004

(a) P_1 (29.4°C at 12:00), P_2 (26.7°C at 15:00) and P_3 (28°C at 21:40) were all pulse temperature increases in Hyderabad on Nov.15, 2004.

(b) In Bangkok, the daily maximum temperature on Nov.15, 2004 (H_1, 36°C) reached its highest on Nov.15 of the 16 years from 1996 to 2012. In Hyderabad, the daily maximum temperature on Nov.15 was as high in 2004 (H_2, 32.2°C) as in 2009 (H_3, 32.2°C), the highest in 16 years between 1997 and 2012.

(c) Black squares A, B and C plot the epicenters of the M6.6 Andaman earthquake, the M7.5 Nicobar earthquake, and the M9.0 Sumatra earthquake on Dec.26, 2004 respectively. Red squares (location names marked), magenta triangles and cyan circles respectively plot airports during Nov.14~17, 2004 having the highest daily maximum temperature on one or more days in records of 1996~2012, one or more pulse temperature increases, and one or more temperatures over 100°C. Black circles have especially few temperature records during Nov.14~17, 2004. Temperature data and earthquake data were from the WU and the USGS respectively.

The Indian Ocean vapor eruptions caused temperature abnormality. Fig.15a shows three pulse temperature increases in Hyderabad on Nov.15. Thus, item 2 "pulse temperature increase" has been satisfied. Furthermore the highest daily temperature on Nov.15 reached its highest in 16 years in Bangkok and Hyderabad (Fig.15b). Thus, item 4 "the daily maximum temperature of a day reaches or surpasses daily maximums of the same day in many years" has been satisfied. Fig.15c plots other airports around the Northern Indian Ocean having abnormal temperatures according to items 2 and 4, and temperature over 100°C in Karachi and Lahore during the eruption on Nov.14~17. This figure also shows many airports lacking temperature data, especially those in Sumatra (Fig. 15c). Temperature data near erupting nozzles are important for finding the nozzles to pinpoint epicenters. They are also important for identifying the second temperature peak following the first temperature peak during vapor eruption, since earthquakes usually occur within a few days after the second eruption and second temperature peak.

Fig.16 shows the second vapor eruptions of the three large Indian earthquakes. The second eruptions were much smaller than the first vapor eruptions (Fig.13), and appeared near the epicenters one or two days before the earthquakes.

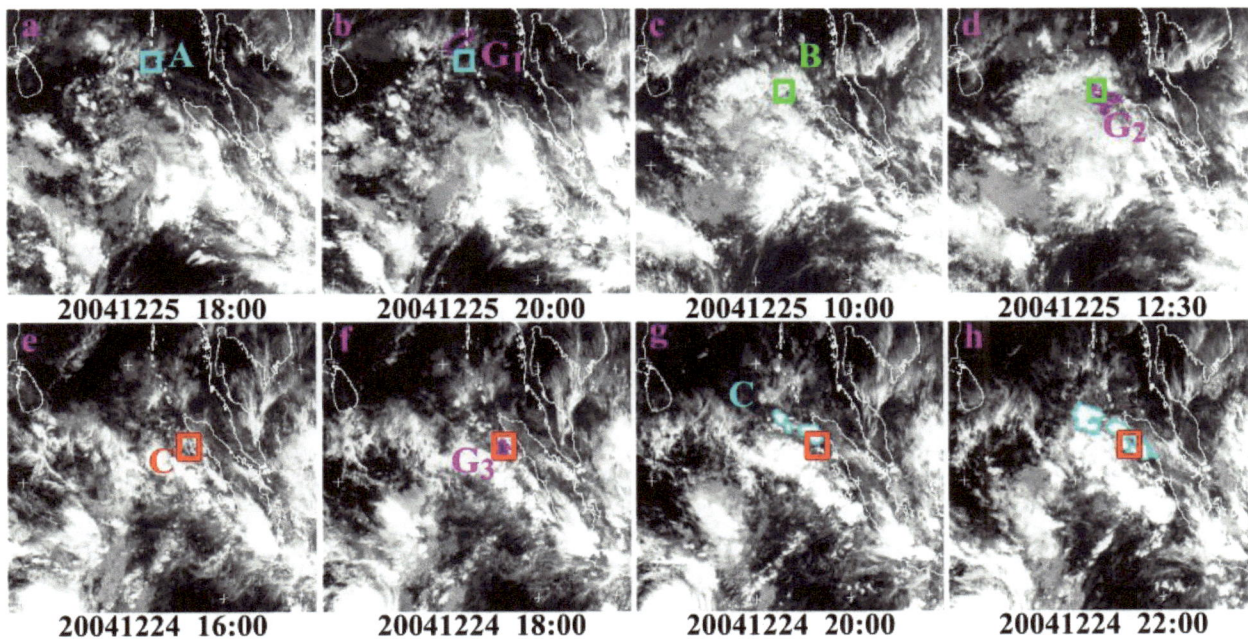

Fig.16: Second eruptions of the Indian earthquakes in 2004
(a) Cyan square A plots the M6.6 Andaman epicenter;
(b) Geoeruption G$_1$ (magenta edge) appeared near A at 20:00 on Dec.25, 2004;
(c) Green square B plots the M7.5 Nicobar epicenter;
(d) Geoeruption G$_2$ (magenta edge) appeared near B at 12:30 on Dec.25;
(e) Red square C plots the M9.0 Sumatra epicenter;.
(f) Geoeruption G$_3$ (magenta edge) appeared near C at 18:00 on Dec.24;
(g-h) Part of vapor rose up to form an earthquake cloud C (cyan edge).

1.9 Summary of earthquake vapors

We have discussed five examples of earthquakes and their vapor eruptions. Vapor eruptions take the form of either earthquake clouds or geoeruptions. The Bam earthquake cloud demonstrates that earthquake vapor indeed erupts from a fixed nozzle of an impending hypocenter. The Northridge earthquake cloud shows an earthquake fog near epicenter. The swarm of Turkey geoeruptions reveals a vapor band between nozzle X and two bulges B and C, under each bulge two earthquakes struck later (Fig.11). These eruptions together depict terms in Fig.3 "Earthquake vapor schematic diagram": earthquake cloud, earthquake fog, geoeruption, nozzle, crevice, band, epicenter, and hypocenter.

The vapor eruptions look very different, but they share common characters. They appear suddenly, have special formations, and cause abnormal temperature increases in their downwind directions, all of which cannot be explained by meteorology.

In contrast, the Earthquake Vapor Model provides a great explanation. It proposes that vapors were generated in corresponding hypocenters. Vapor temperature and pressure gradually accumulate to high levels. When vapor pressure overcomes the resistance in the main crevice, the vapor erupts from a hypocenter, through the main crevice to the nozzle on the surface, and rises up rapidly.

Therefore, the Northridge earthquake cloud looked like a launching rocket, the Bam earthquake cloud persisted in the nozzle for 26 hours, the Kerman earthquake cloud had a length of 370km with a uniform width, and the swarm of Indian Ocean clouds formed three straight lines of about 4,700km in length. Nozzles are usually close to epicenters. In the very rare cases that nozzles are not close to epicenters, minor part of vapor can erupt through small crevices connecting the main crevice to the surface, or directly escape from hypocenter to the surface above. This process forms a vapor band connecting the nozzle and bulges formed above the hypocenters (the Turkey eruption black band XBC in Fig.11). Erupting vapors contain enormous heat, so temperature increases abnormally in their downwind direction. Because hypocenter temperatures reach 300~1520°C, it is understandable why surface temperatures surpass 100°C in Karachi and Lahore during the swarm of Indian Ocean vapor eruption (Fig.13).

The enormous heat carried by vapor eruptions suggests that temperatures of nozzles are similar to those of hypocenters: 300~1520°C. Shou (2011) proposed to precisely measure abnormal temperature increases to pinpoint nozzles where most of earthquakes will hit. In case a vapor band appears between a nozzle and a bulge, an earthquake will hit the bulge (Fig.11). This method will work in a hot surrounding, thus overcoming a limitation of Shou's Bam prediction which relied on a cold surrounding. Moreover, measuring the vapor quantity and comparing it with vapor quantities of previous earthquakes of standardized magnitudes, the magnitude of the impending earthquake can be predicted with high precision.

Shou (2006b) reported that among more than 500 earthquakes, the longest duration from vapor eruption to its associated earthquake was 112 days, with an average of 30 days. How might we narrow the prediction time window?

We have previously discussed that the main vapor eruption of the M6.8 Bam earthquake was not complete on Dec.20, 2003. The remaining water and vapor in the hypocenter erupted again after having accumulated enough energy (Fig.7e), which is associated with the second temperature peak B at Zahedan downwind of the subsequent eruption on Dec.25 (Fig.7c). Afterward, the hypocenter became dehydrated and the Bam earthquake quickly followed (within one day).

The M6.5 Kerman earthquake was another example. The main eruption (Fig.8b~e) and its associated temperature peak A (Fig.8h) occurred on Dec.14, 2004. Subsequent eruptions at the Kerman epicenter (Fig.9b, f~g) caused temperature peak B (Fig.8h) on Feb.20~21, 2005. Temperature peaks in between the two peaks are likely caused by vapor eruptions of other earthquakes near the Kerman epicenter. For example, the M5.5 Chatroud vapor eruption (Fig.10a~d) near Kerman and the temperature peak C in Mashhad downwind from the vapor eruption on Feb.2, 2005 (Fig.8h, 10e) coincided both temporally and spatially, suggesting that the M5.5 Chatroud eruption caused temperature peak C. Therefore, isolating a nozzle to measure its second temperature peak can automatically predict an impending earthquake from that nozzle within a few days.

Based on these work, Shou (2011) proposed a method to reduce the time window of an earthquake prediction. He pointed out that with 10% probability, an earthquake would occur within the first three days after its vapor eruption. For example, just 17 hours after an earthquake cloud on Jun.20, 1990, the M7.7 Northern Iran earthquake struck (Shou, 1999). This phenomenon may be because this eruption was complete. With 90% probability, an earthquake would occur within a few days after its second eruption or second temperature peak. For example, the Northridge earthquake (Fig.5d), the Bam earthquake (Fig.7c, e~g), the Kerman earthquake (Fig.8h, 9b~c, f~g), the moderate Turkey earthquakes (Fig.12b), and the three large Indian Ocean earthquakes (Fig.16) all happened within a few days after the second eruption or second temperature peak. Combining both possibilities, we can narrow a time window into a week.

The vapor eruptions of the Northridge earthquake, the Bam earthquake, and the Kerman earthquake happened where no big earthquake had struck in recorded history; while vapor eruptions of the Indian Ocean earthquakes happened where many big earthquakes had struck. The Indian Ocean vapor eruptions occurred in ocean, while other examples were in land. The Turkey geoeruptions happened in depths less than 10km, while other examples happened in depths more than 10km. These examples show that the Earthquake Vapor Model is universal.

1.10 Discussion

The earthquake vapor model has met with skepticism. The main question was about whether or not the vapor can make surface temperature surpass 100°C because such temperature would kill anybody nearby. Indeed the vapor of such high temperature would kill anyone nearby, but fortunately, such case is rare. There is usually a distance between a nozzle and people. Moreover, hot erupting vapor always rises up while cold air sinks down, forming a convection.

Thus temperature at a location near a nozzle can decrease sometimes (see Fig.5b and Fig.8g). In addition, there is a big difference between 100°C pure vapor and 100°C air. The former can burn someone, while the latter may not because of its low thermal capacity and conduction. After passing through a distance, the vapor is diluted with air and may not kill everybody nearby. On the other hand, very hot erupting matter burnt a man during the Tangshan earthquake on Jul.28, 1976 (Shi et al., 1980), hibernating snakes crawled out from their holes and died on frozen ground before the Haicheng quake on Feb.4, 1975 (Jiang and Du, 1984).

Maybe because surface temperature above 100°C is unimaginable, such data are often hidden, skipped, tampered or misunderstanding. For example, the GOES-12 Image maps scene brightness to temperatures between "0 and 342.096K", where 342°K equals 69°C. An image may contain an area of fires with temperature sometimes exceeding 342K@12 Temperature above 69°C will be hidden. Under such scale, there would be no big difference between a nozzle and its vicinity in satellite imagery, which affects earthquake prediction seriously.

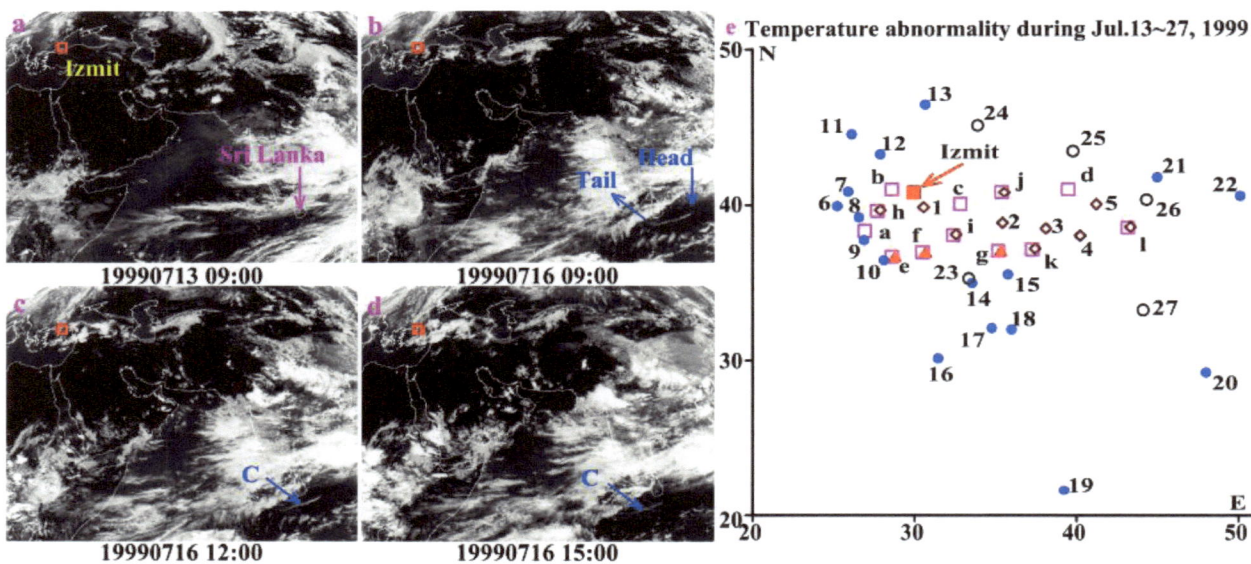

Fig.17. Turkey Vapor Eruption and Temperature Data Loss in July 1999

(a) Red square plots Izmit (40.75, 29.86) where an M7.7 earthquake struck on Aug.17, 1999.

(b) A linear cloud (C) in clear sky near Sri Lanka was moving northeastward against an opposite wind at 9:00 on Jul.16. It had a length of 800km indicating a magnitude above 7. Its tail pointed toward northwest (cyan arrow), an area from Iran to Italy.

(c-d) The cloud became shorter and was pushed south-west direction because of the opposite wind, and disappeared after 15:00.

(e) The red solid square plots the M7.7 Izmit epicenter. Others with letters or numbers are airports, whose names, countries, coordinates, temperatures and states are in Table 3. Magenta hollow squares, brown hollow diamonds, and red solid triangles plot Turkish airports that lost many temperature data on Jul.13, lost many temperature data on Jul.14 and no any data on Jul.15-17, and had the highest daily maximum on Jul.14, 1999 in 15 years from 1996 to 2010 respectively. Blue solid circles and black hollow circles plot airports out of Turkey that did not lose data and do not customarily record temperature data, respectively. Satellite images and temperature data were from DU and the WU respectively.

@12 http://www.oso.noaa.gov/goes/goes-calibration/G12_Img_Ch2_Rollover/G12_Ch2_Rollover_Abs.pdf

For instance, Shou (1999) found a linear earthquake cloud near Sri Lanka on Jul.16, 1999. According to its length of 800km, its tail pointing toward northwest, and its appearance time on Jul.16, he predicted an earthquake of magnitude 7 or above from Iran to Italy within 34 days to three witnesses on Jul.30. He tried to find a trace from the tail of the cloud to the impending epicenter, but failed. On Aug.17, an M7.7 earthquake hit Izmit, Turkey (between Iran and Italy). This earthquake was the only one of magnitude 7 or above in the predicted area within 915 days. Fig.17a~d show satellite images from Izmit (red square) to Sri Lanka on Jul.13~16, 1999. There is no difference in darkness (corresponding to high temperature) between Izmit and its vicinity, so the tail of the cloud could not be traced to Izmit.

After the Izmit earthquake, many Turkish scientists and people wrote Shou that the weather there had been extremely hot even with air conditioning, and was the hottest within at least 60 years. However, Shou did not understand why he could not find extremely high temperature data until realizing that temperature data had been skipped (Shou, 2011). Fig.17e and Table 3 depict the severity of missing data. On Jul.13, 1999, twelve Turkish airports, including Istanbul and Ankara, lacked many hourly temperature records. On Jul.14, 1999, three Turkish airports got the highest daily maximum on Jul.14 in 15 years from 1996 to 2010. On Jul.15~27, ten Turkish airports recorded nothing. By contrast, airports out of Turkey recorded temperature data. The coincidence between missing temperature data and the Izmit earthquake vapor eruption in time and space suggest that extremely high temperature data were probably too unimaginable to be recorded

The Weather Underground (WU) website displayed many surface temperature data above 60°C, recorded by airports. Table 4 reveals 18 examples of abnormal temperatures preceding earthquakes. Those data have all been deleted in the WU website, after the website replied on Jul.23, 2010 that "we assume that any temperature greater than the highest temperature reliably measured on Earth (136F, which is 59°C) is an error, and should be deleted". It is well known that modern thermometers can precisely measure temperature higher than 100°C, but weather stations and major airports are accustomed to skip high temperature. A small number of airports reported high temperature, which are important evidence supporting Shou's theory (although these airports do not dare to announce these high temperatures as daily maximums). Surface temperature data above 100°C are important. Artificially deleting those rare data is an example of tampering with scientific data.

The NCDC has been establishing a vast world weather database. We have cited its data previously to verify the Earthquake Vapor Model. This database has a big problem: its data are not standardized. Different stations have different data frequency. Data losses widely exist even among good stations and good airports. For instance, the Sandberg station skipped temperature data at 17:08 and 17:28 (two hours after the Northridge cloud) and at 22:08 (twenty minutes before the station got its claimed peak temperature) on Jan.8, 1994 (see Fig.5c).

Fig.18a~b reveals a comparison of temperatures between the Kerman airport and the Kerman station. The former recorded a pulse of 141°C at 14:20 (LT 18:20) on Dec.15, 2004 (Fig.18a, A) and another pulse of 24°C at 16:20 (LT 20:20) on Dec.20, 2003 (Fig.18b, B), while the latter skipped the both. The airport also made skips at 19:50 (Fig.18b, magenta "+") on Dec.20. The station and the airport made two skips: one at 12:50 and the other at 2:50 on Dec.15 (Fig.18a, S_2 and S_3). To solve this problem, the NCDC might need to set up a common frequency of at least 96 data per day and urge all stations and airports to record all data. Furthermore, stations do not distribute evenly and densely. It is very difficult to find data at or near nozzles.

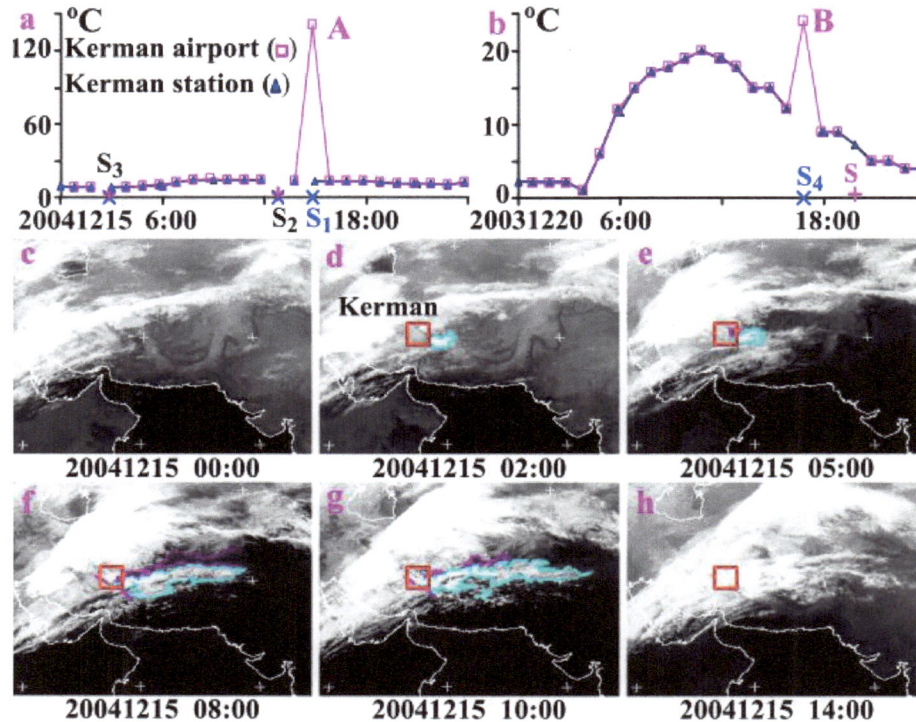

Fig.18. Temperature data skips and corresponding vapor eruption

(**a**) The Kerman airport (magenta square) recorded a pulse temperature A from 13°C at 13:20 (LT 17:20) to 141°C at 14:20 (LT 18:20) on Dec.15, 2004; while the Kerman station (blue triangle) skipped it (S_1). Both made two skips S_2 at 12:50 and S_3 at 2:50.

(**b**) The Kerman airport (magenta) recorded a pulse temperature B from 12°C at 15:20 (LT 19:20) to 24°C at 16:20 (LT 20:20) on Dec.20, 2003 during the eruption of the Bam earthquake vapor, while the Kerman station (blue) skipped this point (S_4) and the Kerman airport skipped 19:50 (S).

(**c~h**) One day after the M6.5 Kerman earthquake cloud (Fig.8), earthquake clouds (Panel d, Cyan edge) appeared near Kerman at 2:00 on Dec.15, 2004. The clouds grew in size with geoeruption (magenta edge) at 5:00~10:00 (Panel e~g), and overlaid Kerman at 15:00 (Panel h), coinciding with the pulse temperature of 141°C (Panel a). These clouds corresponded to a swarm of moderate earthquakes on Feb. 22, 2005, the same day as the M6.5 Kerman earthquake. Thus, the swarm of earthquakes should not be considered as aftershocks as the M6.5 Kerman earthquake. Temperature data, earthquake data and images were respectively from the WU, the USGS and DU.

Some scientists questioned whether a vapor bubble could rise from the sea bed of the Atlantic Ocean through a depth of 10km to the surface. This question is easy to answer because a vapor bubble is much lighter than the weight of equal-volume water. In fact, Harrington and Shou (2005) had shown that the vapors of the M7.1 off coast of Northern California earthquake on Sep.1, 1994, the M6.3 off coast of Oregon earthquake on Oct.27, 1994 and the M6.8 off coast of

Northern California earthquake on Feb. 19, 1995 (corresponding to *Figs.4.3, 4.4 and 4.5* or Figs.27.c~e in this book respectively) originated from depths of 10~20km. Shou (2006b) had also revealed the vapors of the three large Indian Ocean earthquakes on Dec.26, 2004 that were all deeper than 15km. Two of them were deeper than 30km.

We will discuss an important question: how much pressure an impending hypocenter may have. Shou (2006b) estimated a minimum pressure of 1532 atm by the existence of vapor cloud and depth of the Sumatra tsunami earthquake (16.1km). Later, the USGS increased the depth to 30km, so the pressure should correspondingly increase to about 3000 atm (1 atm pressure is generated by about 10m of water). The real pressure should be much greater to form the straight line of 4700km (CZ of Fig.13j~k).

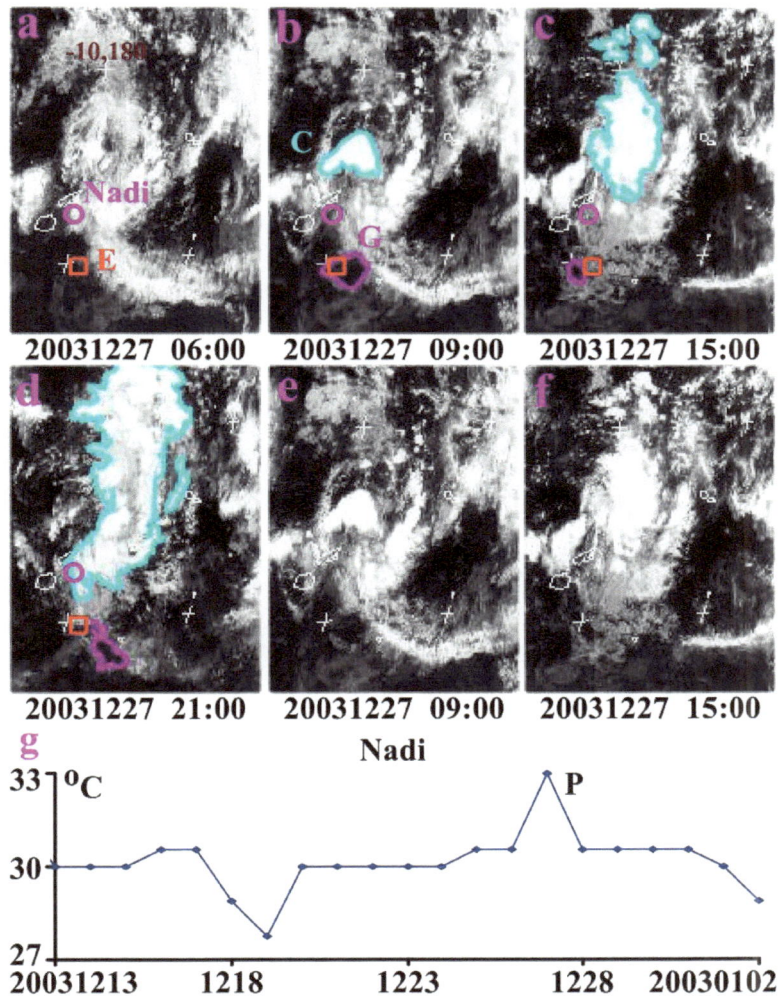

Fig.19. The M6 Fiji Earthquake Cloud
(**a**) Red square E and magenta circle respectively plot the M6 Fiji epicenter (-20.14, -179.15) on Jan.11, 2004 and the Nadi airport (-17.8, 117.4). The earthquake had a depth of 673.1km. Brown numbers (-10, 180) note the coordinate of a white plus northward from Nadi.

(**b-d**) A geoeruption G (magenta) near the epicenter and an earthquake cloud C (cyan) appeared at 9:00 on Dec.27, 2003. The cloud became bigger and moved northeastward later.

(**e-f**) Unmarked versions of Panel **b**~**c** show the exactly initial epicenter.

(**g**) The Nadi airport recorded an increase of daily maximum temperature from 30.6°C on Dec.26, 2003 to 33°C (Point P) on Dec.27, coinciding with the cloud. The temperature returned back to 30.6°C shortly afterward. Temperature data. earthquake data and images were respectively from the WU, the USGS and DU.

Among big earthquakes in 1990~2012, the deepest depth is 673.1km of the M6 Fiji earthquake (-20.14, -170.15) on Jan.11, 2004. Fig.19 reveals its main cloud C (cyan), emerging at 9:00 on Dec.27, 2003, moving northward and becoming bigger at 15:00~21:00. Meanwhile, a small geoeruption G (magenta) appeared near the epicenter (red square). In the erupting direction, the Nadi airport recorded an increase of daily maximum temperature from 30.6°C on Dec.26 to 33°C on Dec.27 (P of Fig.19g), coinciding with the appearance of the cloud. The cloud expanded at least 1466km from the epicenter. Similar to the Sumatra tsunami earthquake, this hypocenter can reach a pressure of **67055** atm at least ($=1+\rho gh/ 101325 =1+1030$ (kg/m^3) x9.8(m/s^2) x673100(m) /101325).

If both Sumatra earthquake and Fiji earthquake have equal pressure, then the pressure for vapor to move 1km should be 19.8 (atm) $=(67000-3000)/(4700-1466)$. Thus, the hypocenter pressure may reach **96000** atm ($=67000+19.8$x 1466 or 3000+19.8x4700). Of course, it is better to verify the ratio of 19.8 atm/km by experiment.

Chapter 2 Earthquake Prediction by vapor

We have discussed the earthquake vapor model and several types of earthquake vapor appearances. To better understand the model, we will discuss how nozzle structure and the surrounding environment including temperature, wind, cloud, and topography might affect the vapor. We will discuss how to predict earthquakes and challenges associated with predictions. We will also discuss whether or not earthquakes can occur without prior vapor eruptions or vapor eruptions are not followed by any earthquake for a long period of time (four months). We will discuss how problems of satellite data, earthquake data, and the lack of empirical data can affect predictions.

2.1 Various appearances of geoeruptions

We have already discussed several geoeruptions in Figs.7~11, 13, 16, 18 and 19 in Chapter 1. Here, we will discuss more examples.

2.1.1 Belt-shaped geoeruption

Fig.20 shows three belt-shaped geoeruptions originating from impending epicenters. On Aug.14, 1999, geoeruption G_1 appeared in Bolinas Bay, Northern California southwestward (Panel a). Shou did not check current data to make his first prediction by geoeruption (Table 10 No.37) on Aug.25. However, an earthquake of magnitude 5 had hit Bolinas Bay on Aug.18, plotted by red square E_1 coinciding with the origin of G_1. This earthquake was the only one of magnitude ≥ 5 in an area of $\pm 2°$ around the epicenter within 752 days from Aug.13, 1998 to Sep.2, 2000. A smaller geoeruption G (arrow) had its tail pointing toward an M4.3 epicenter E (cyan filled red square).

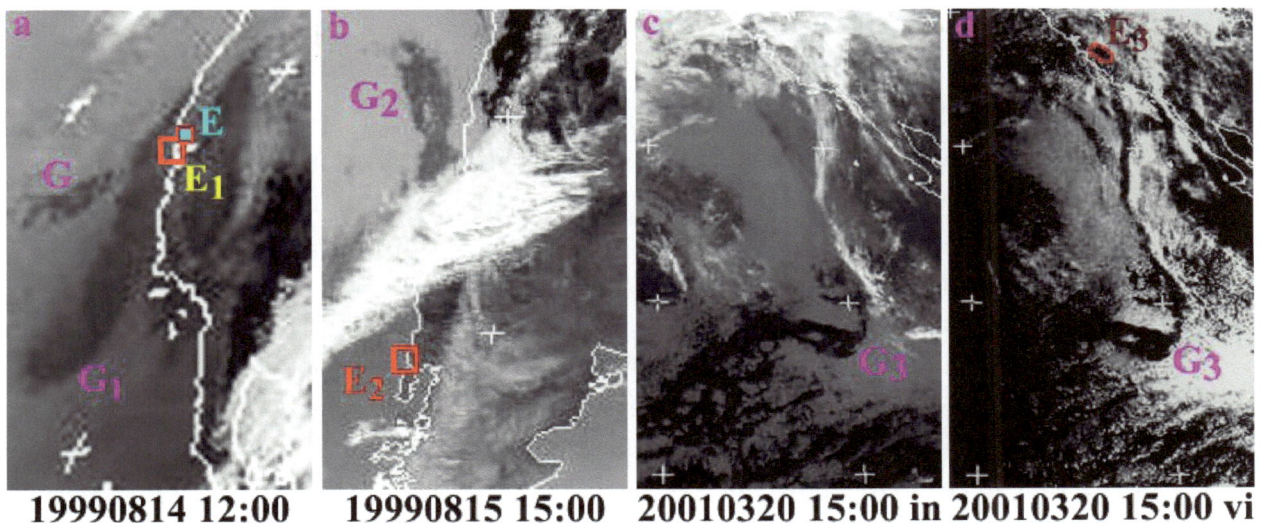

19990814 12:00   19990815 15:00   20010320 15:00 in 20010320 15:00 vi

Fig.20 Belt -shaped geoeruptions

(**a**) E_1 and E (red squares) respectively plot the epicenters of the M5 Northern California earthquake at (37.9, -122.68) on Aug.18 and the M4.3 at (38.39, -122.63) on Sep.22, 1999. From E_1 and E, goeruptions G_1 and G moved southwestward on Aug.14.

(**b**) Red square E_2 plots the epicenter of the M6.4 earthquake (-40.5, -74.75) off coast of Chile on Aug.22, 1999. From E_2, geoeruption G_2 moved northwestward on Aug.15.

(**c-d**) Fig.20**c** and **d** present a comparison between an infrared image and a visible image in the same area of Southern California at 15:00 on Mar.20, 2001 when geoeruption G_3 was erupting from E_3 (Red circle, Plane d) southward and then turned to west. The infrared image did not show G_3 completely, while the visible image did clearly. The red circle was made by Shou to predict an earthquake of magnitude ≥4 within the circle from Apr.3 to Jul.2 to the USGS (Appendix9, Table 11 No.57) and the public. On Jul.2~3, a triplet of M4 earthquakes (36.7, -121.3) hit Hollister in the circle. The images and earthquake data were from DU and the USGS respectively. All images except for Fig.20d are infrared.

Geoeruption G_2 was moving off coast of Chile northwestward on Aug.15, 1999 (panel b). On Aug.22, an M6.4 earthquake hit the nozzle of G_2 (red square E_2) at (-40.5, -74.75). This earthquake was the only one of magnitude ≥6.4 in an area of ±20° around the epicenter within 376 days from Sep.4, 1998 to Sep.14, 1999. The high spatial coincidence between the nozzle of the geoeruption and the earthquake, and the fact that the geoeruption maintained its non-meteorological shape instead of being dispersed suggest that during eruption from the impending epicenters, wind force was weak.

On Mar.20, 2001, geoeruption G_3 (Fig.20c~d) was moving from Hollister, Southern California (red circle E_3, Fig.20d) southward. Shou drew this circle to predict to the USGS (Appendix9) and the public an earthquake of magnitude ≥4 within the circle from Apr.3 to Jul.2. On Jul.2~3, a triplet of M4 earthquakes hit the circle exactly. The triplets were the only triplets of earthquakes of magnitude ≥4 in an area of ±2° around the epicenter within 4995 days from Apr.19, 1990 to Dec.21, 2003. The visible image (Fig.20d) showed the circle clearly, while the infrared image (Fig.20c) did not. This comparison reveals that earthquake vapor appears differently in satellite images taken at different wave length.

2.1.2 Earthquake clouds and geoeruptions can appear as different darkness at different wavelength

Fig.21 describes another comparison between visible and thermal infrared images on earthquake clouds and geoeruptions of the same area of Japan and Russia, and at the same time of 3:00 on Aug.20, 2003. Geoeruption Gv and earthquake clouds Cv's in visible channel (Fig21a) are much clearer than the corresponding Gi and Ci's in infrared channel (Fig21b).

Visible light has a wave length of 0.5~0.7μm, and thermal-infrared light has a wave length of 10.3~11.3μm. Geoeruption is much warmer than earthquake clouds, so the average distance between vapor molecules is larger in geoeruption, probably greater than 0.7μm. Thus visible light but not infrared light will all pass through geoeruption and be absorbed by the Pacific Ocean below instead of being reflected back to the satellite camera. Therefore, Gv appeared much darker than Gi.

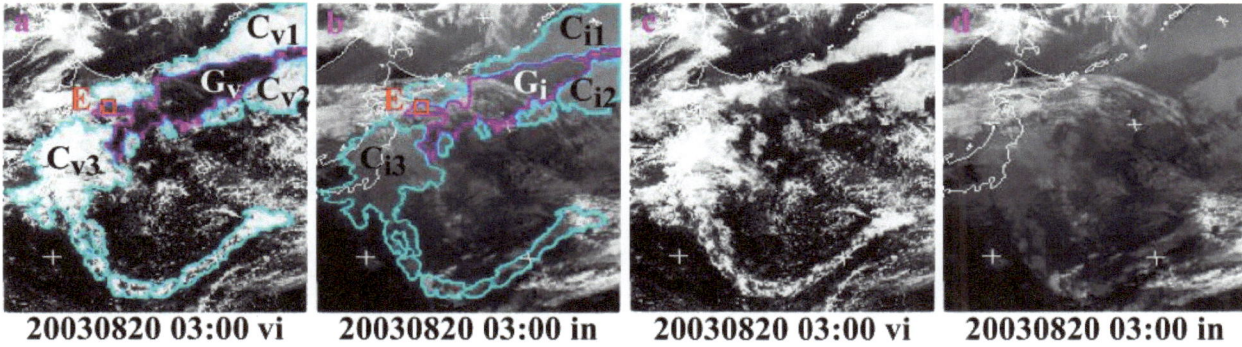

20030820 03:00 vi   20030820 03:00 in   20030820 03:00 vi   20030820 03:00 in

Fig.21 Vapor appearance at different wavelengths

(a-d) The four images show the same area in Japan and Russia at the same time: 3:00 on Aug.20, 2003. Panels **a** and **c** are visible; while **b** and **d** are infrared. Panels **c** and **d** are unmarked original. Red square E plots the M8.3 epicenter (41.81, 143.91) on Sep.25, 2003. Magenta and cyan outline geoeruption and earthquake clouds respectively. The images were from DU. The earthquake data were from the USGS.

The average distance between vapor molecules of the clouds should be less than 0.7μm, so visible light emitted from satellite will be reflected back to and received by satellite. A cloud will reflect visible light with a chance of about 20 (=10.3μm/0.5μm) times higher than reflecting thermal-infrared light. Thus, earthquake clouds Cv (Fig21a) looks white, while earthquake clouds Ci (Fig21b) looks gray. When a cloud is cold and heavy, Cv and Ci can have same color. The vapor from the impending epicenter E erupted northeastward to form a geoeruption, and both sides of the geoeruption became earthquake clouds.

2.1.3 Spot-shaped geoeruptions

Fig.22 reveals several spot-shaped geoeruptions in Taiwan on Jan.30, 2000. Over the subsequent 46 days, a series of eight earthquakes occurred at exactly the locations of the dark spots (Table 5). The average absolute errors between the earthquake and the geoeruption point of origin are 0.09° for latitude and 0.15° for longitude. The high spatial coincidence between the geoeruptions and the earthquakes suggests calm weather accompanying

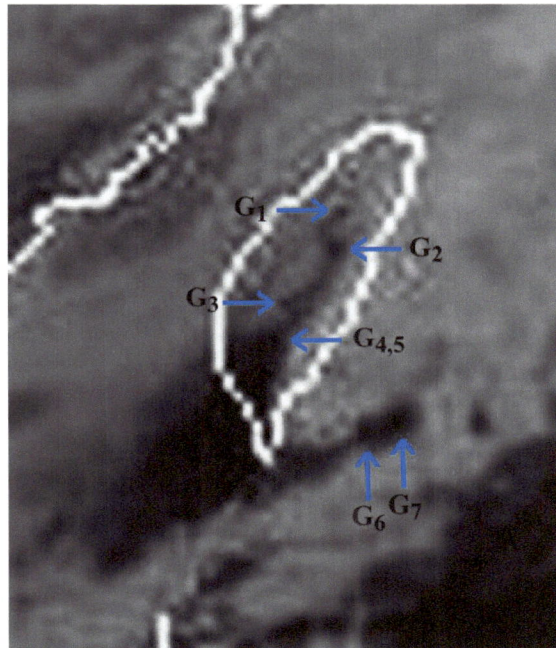

Fig.22 Taiwan spot-shaped geoeruptions

Several dark spots (arrows), indicating warm regions, appeared in the midst of cloud cover at 3:00 on Jan.30, 2000. They were geoeruptions. Shou predicted earthquakes with this image to the public immediately. Over the next 46 days, a series of eight earthquakes occurred at exactly the locations of these dark spots. The earthquake data are shown in Table 5. The pixilated appearance of the image is a result of magnifying a small jpeg file. The image was from DU.

these geoeruptions. This high coincidence also suggests the vapor precursor being exact. Therefore, the vapor precursor may be used to identify missing earthquake data. Missing earthquake data in earthquake catalogues are common: among eight earthquakes, the USGS and the Central Weather Bureau of Taiwan (CWBT) recorded 4 and 6 earthquakes, respectively (Table 5). Table 5 further reveals that the location data of earthquakes from both USGS and CWBT are quite precise, but the data on depth can have big discrepancy.

Table 5 Taiwan geoeruptions vs. earthquakes reported by different sources

Geoeruptions					Earthquakes							
Date UTC	Time	P	Lat. N	Lon. E	Date UTC	Time	Lat. N	Lon. E	Mag. ML	mb	Dep. Km	S
20000130	3:00	1	24.4	121.1	000131	21:11	24.37	120.9	4.6		4.2	T
					000216	19:48	24.35	120.8	4.0		7.4	T
		2	24.0	121.2	000130	20:21	23.90	121.31	4.8	4.1	33	U
							23.90	121.31	4.8		7.5	T
		3	23.5	120.7	000131	2:57	23.51	120.48	4.2		4.7	T
		4	23.2	120.7	000215	21:33	23.35	120.93		5.3	33	U
							23.33	120.75	5.6		21.1	T
		5	23.2	120.7	000216	0:33	23.33	120.75	4.5		13.4	T
		6	22.2	121.4	000226	8:23	22.24	121.37		4.1	33	U
		7	22.2	121.8	000316	0:37	22.06	121.62	5.0	4.8	33	U

Note:

1. P: point number in Fig.22. Lat. latitude. Lon. longitude. Mag. magnitude. Dep. depth. S: source. U: the USGS. T: the CWBT (Central Weather Bureau of Taiwan) @13.

2. The latitudes and longitudes of the geoeruptions were calculated directly from the image, and have an uncertainty of 0.2°.

3. The average latitude and longitude absolute errors between the earthquakes and the geoeruption points of origin are 0.09° and 0.15°, respectively.

@13 http://www.cwb.gov.tw/V7/index.htm

2.1.4 Hurricane-linked geoeruption

Fig.23 depicts a hurricane-linked geoeruption above the Dominican Republic on Aug.22~23, 2003. Erupting earthquake vapor from an impending M6.5 epicenter E at (19.68, -70.67) dissipated part of a moving hurricane cloud. This formed a rare deformation in the hurricane cloud from 3:00 on Aug.22 to 9:00 on Aug.23. When the hurricane cloud fully covered Dominican Republic at 21:00 (LT 17:00, Fig.23d) on Aug.22, the Puerto Plata airport, 15 km away from the epicenter E, recorded a sharp pulse temperature increase of $10^{o}C$ within 10 minuets from $28^{o}C$ at 20:50 (LT 16:50, Fig.23c) to $38^{o}C$ at 21:00 (LT 17:00, Fig.23d). This temperature was the highest daily maximum from Aug.1 to Sep.30.

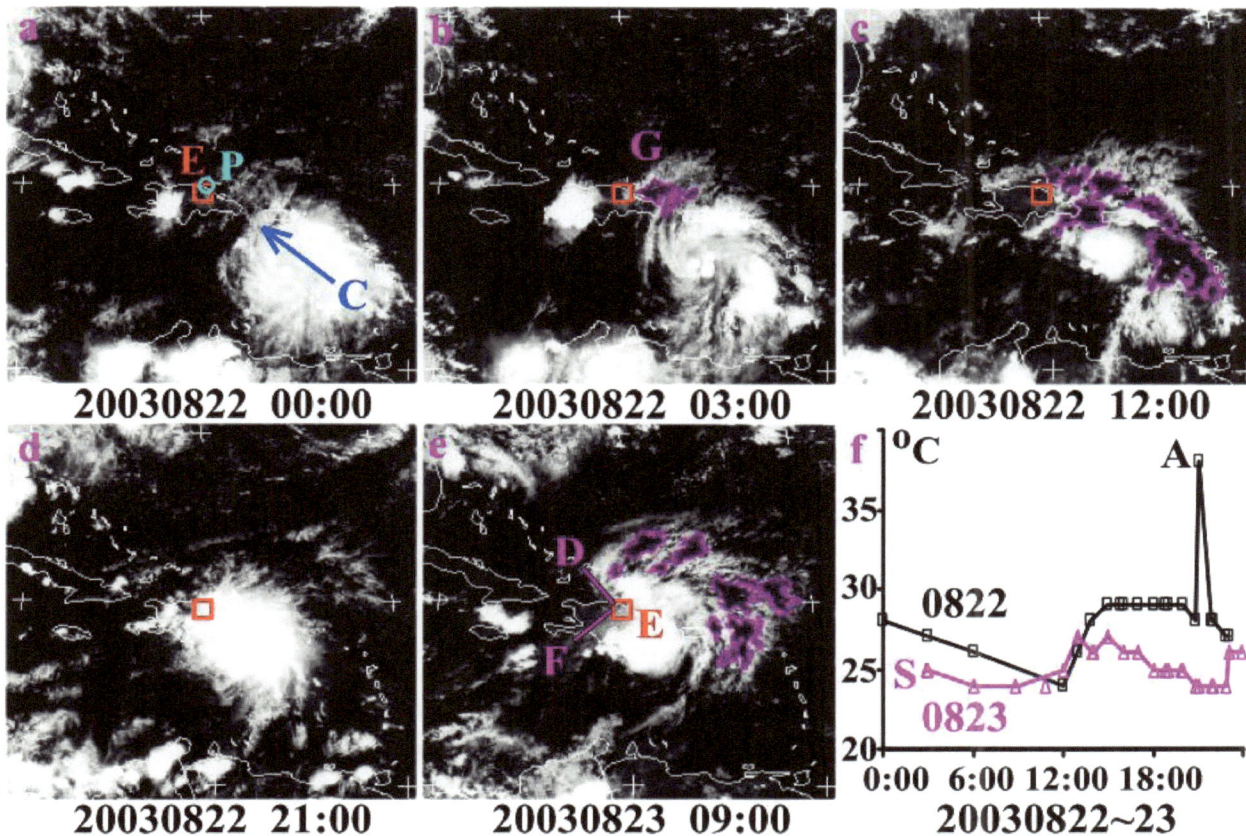

Fig.23 Hurricane-linked generation

(**a**) Red square E and cyan circle P respectively plot an M6.5 epicenter at (19.68, -70.67) in Dominican Republic on Sep.22, 2003 and the Puerto Plata airport (19.8, -70.6), 15 km away from epicenter E. Hurricane cloud C was moving counterclockwise at 0:00 on Aug.22 northwestward (Arrow).

(**b-c**) Geoeruption G (Magenta) formed when the hurricane cloud C encountered erupting vapor from E at 3:00. Geoeruption became bigger at 12:00, changed the circle outline of hurricane cloud.

(**d-e**) The hurricane cloud C covered E at 21:00, and at 9:00 on Aug.23, formed an angle DEF whose tip coincided the M6.5 epicenter E.

(**f**) When cloud C covered epicenter E at 21:00 on Aug.22, the Puerto Plata airport recorded a pulse temperature increase from 28°C at 20:50 (LT 16:50) to 38°C at 21:00 (LT 17:00) within 10 min. This temperature pulse was absent on other days, e.g. Aug.23 when skipped datum S.

2.1.5 Topography-linked geoeruption

Fig.24 presents the identical Iran geoeruption on Dec.15, 2004 shown in Fig.18. The shape of this geoeruption was influenced by topography. Erupting vapor from Kerman formed a strong geoeruption G eastward, a fixed geoeruption MIJ, and mixed clouds C_1 and C_2 consisting of earthquake clouds and weather clouds from the west (Fig.24b-d). A high spatial coincidence between mountains Masahun–Ilazaran-Jebal Barez (M-I-J, Fig.24e) and geoeruption MIJ above the mountains suggests the vapor rushing to and then rising up along mountain range M-I-J to form geoeruption MIJ, and therefore MIJ and M-I-J had similar shape and locations. Therefore, MIJ was topography influenced by the mountain range M-I-J. Similarly, G was topography influenced by the mountain Kuhpayeh.

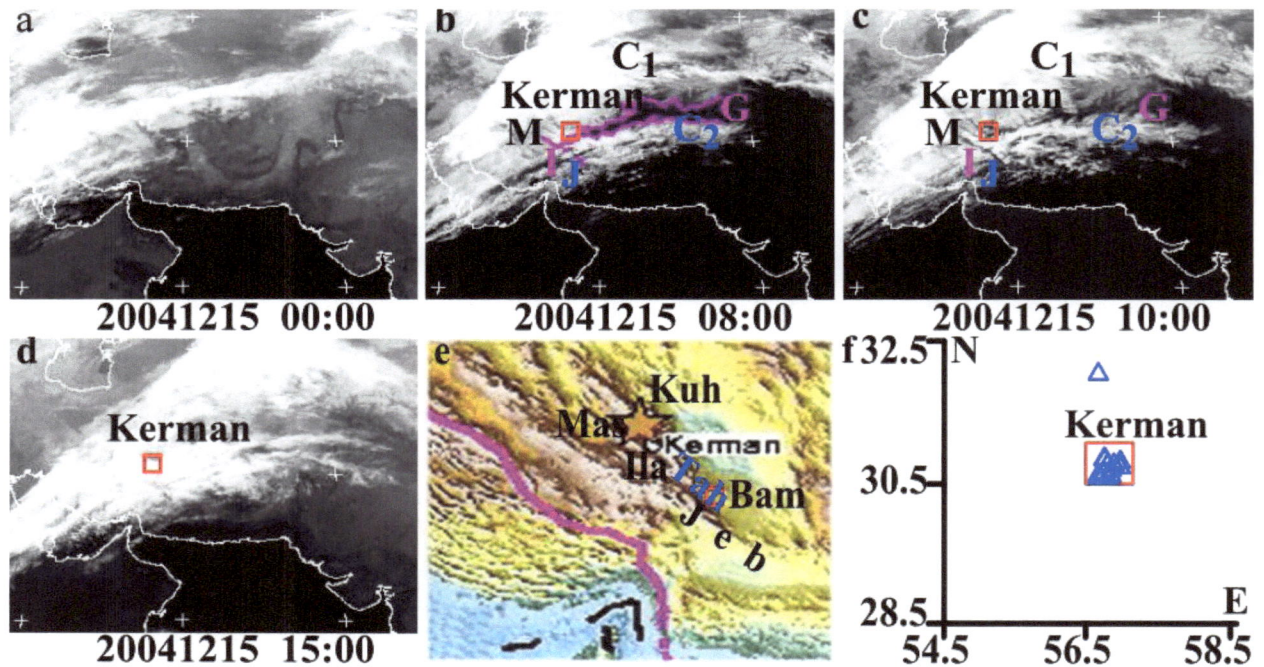

Fig.24 Kerman topography-linked geoeruption
(**a-c**) Vapor erupted from Kerman (red square) to form geoeruption G, and a fixed linear geoeruption MIJ (magenta outlines in **b**), and mixed with weather cloud from the west to form clouds C_1 and C_2 at 8:00-10:00 on Dec.15, 2004.
(**d**) Both G and MIJ disappeared and mixed clouds became heavy at 15:00 when the Kerman airport recorded a peak temperature of $141°C$ at 14:20 (LT 18:20, Fig.18a, Link 15 of Table 4).
(**e**) Kerman has mountains including Masahun (3600m), Ilazaran (4420m), Jebal Barez in the southwest, and Mountain Kuhpayeh (3142m) in the Northeast. Lake Tahrud is between mountains and near Kerman-Bam highway. Geoeruption MIJ coincided with mountains Masahun–Ilazaran–Jebal Barez.
(**f**) Red square E and blue triangles plot the M6.5 Kerman earthquake (30.74, 56.83) and all 12 moderate shocks in an area of $±2°$ around the epicenter and within 112 days after the M6.5 earthquake on Feb.22, 2005. Eleven shocks were within 23km to the M6.5 earthquake. Satellite images were from DU. Earthquake data and the part of earthquake map were from the USGS.

One may ask how vapor could reach Mountain Jebal Barez from Kerman. There was a highway along Lake Tahrud (Fig.24e) and its connected river from Kerman to Bam. This highway and the lake and the river had lower altitude to offer the vapor a way to reach Mountain Jebal Barez. When lots of vapors erupted and formed heavy clouds above Kerman at 15:00 (Fig.24d), the Kerman airport recorded a temperature of $141°C$ at 14:20 (LT 18:20, Fig.18a, Link 15 of Table 4).

In Chapter 1, we have discussed a linear cloud of 370km in length on Dec.14, 2004 (Fig.8, one day before the geoeruption of Fig.24). This cloud has already predicted the M6.5 Kerman earthquake (19.68, 70.67) on Feb.22, 2005 because of its length. The geoeruption on Dec.15, 2004 likely corresponds to a swarm of eleven moderate earthquakes within 23km of and shortly after the M6.5 Kerman earthquake (Fig.24f). Thus, conventionally regarded aftershocks also have vapor eruptions.

2.2 Various appearances of earthquake clouds

We have discussed the Northridge cloud (Fig.5), the Bam cloud (Fig.6), the second Bam cloud with geoeruption (Fig.7), the Kerman cloud with geoeruption (Fig.8), the second Kerman cloud with geoeruption (Fig.9), the Chatroud cloud with geoeruption (Fig.10), a swarm of Indian clouds with geoeruptions (Fig.13), the second Indian clouds with geoeruptions (Fig.16), the Turkey cloud (Fig.17), a swarm of Kerman clouds with geoeruptions (Fig.18), and the Fiji cloud with geoeruption (Fig.19). We will discuss more types of earthquake cloud appearances.

When making his Bam earthquake prediction, Shou estimated a mountain range near the nozzle which shaped the direction of the cloud. Since the mountain range could be a fault, Shou extended the predicted area from a small circle around the nozzle to fault AB (Fig.7a). After seeing the Bam earthquake map of the USGS, he found that his initial hypothesis had been correct: Jebal Barez (Fig. 24e) was the mountain range along which the Bam vapor had risen up. Since Jebal Barez is not a fault, the nozzle predicts exact the epicenter. He also found Lake Tahrud (Fig.24e) and its connected river near Kerman-Bam highway by which the Bam vapor had passed through to form a pulse temperature of 24°C in Kerman (Fig.14a, 18b).

2.2.1 Topography-linked earthquake cloud

Fig.25 reveals two earthquake clouds above Alberta, Canada on Jul.9~10, 2001. Shou thought that both clouds could have come from two possible areas: the Rocky Mountain of Alberta or the distant East Pacific Ocean. Even though clouds C_1 and C_2 seemed to have originated near the Rocky Mountain, their vapors could have come from East Pacific Ocean, risen up along the mountain (elevation 4,400m), lost major of its heat, and condensed into clouds. This would make these clouds look as if they had emerged from the mountain. The length of the longer cloud (about 330km for C_1) predicted a magnitude ≥6. Thus on Jul.16, Shou predicted to the USGS an earthquake of magnitude ≥6 in South Alberta or its west neighboring area in Canada (<53, -120 ~-112) within 75 days (Table 11 No.61).

Afterward on Jul.31, Shou modified the prediction area to (42~53, <-112) to include a larger area of the Pacific (Table 10 D23). An M6 earthquake hit the ocean (48.69, -128.71) on Sep.14, 2001. It was the only earthquake of magnitude ≥6 in the area of (42~53, -135~-112) within 245 days from Jan.12 to Sep.13, 2001. On the other hand, no earthquake hit Alberta. Fig.25e shows daily maximum temperature in Vancouver from Jul.1 to Sep.14, 2001. Temperature peak of 25.6°C (Peak A) happened on Jul.10 when cloud C_1 reached its longest length. This temperature supported the hypothesis that the vapor had come from the ocean. Like other earthquakes, peak B appeared a few days before the earthquake.

This example also depicts a satellite data problem. Because area hotter than 69°C is represented with the same shade of black, these images do not distinguish the hot epicenter from its warm surrounding. Shou's modified prediction is excellent, but we will choose the wrong initial one to follow a later requirement of evaluation of the USGS.

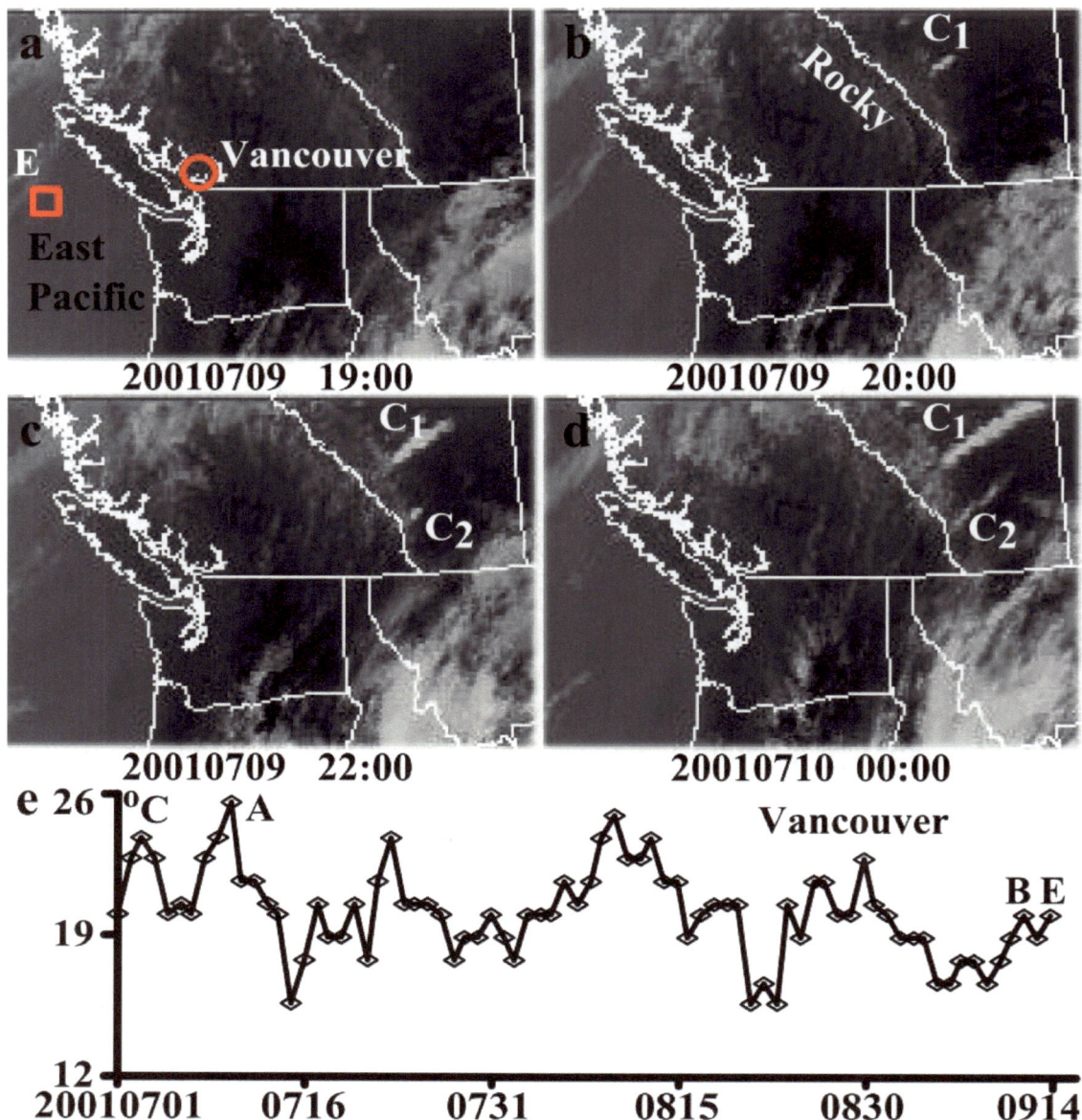

Fig.25 Canada topography-linked cloud

(**a**) Red square and circle plot the M6 earthquake at (48.69, -128.71) off Coast of Vancouver on Sep.14, 2001 and the Vancouver airport (49.2, -123.2) respectively.

(**b-c**) Earthquake clouds C_1 and C_2 emerged above the Rocky Mountain at 20:00 and 22:00 on Jul.9, 2001 respectively.

(**d**) Two clouds became bigger at 0:00 on Jul.10. The length of cloud C_1 increased to about 330km.

(**e**) 'A' reveals a peak in the daily maximum temperature in Vancouver on Jul.10. It was the highest from at least Jul.1 to Sep.14. 'B' reveals a temperature peak two days before the date of earthquake (E). Satellite images and temperature data were from NOAA@14 and WU respectively.

@14 http://www.goes.noaa.gov/

- 38 -

2.2.2 Wave-shaped earthquake cloud

Fig.26a is a schematic diagram for the formation of a wave-shaped earthquake cloud. A wave-shaped earthquake cloud forms when earthquake vapor erupts from a row of nozzles and a wind blows vapors perpendicularly to the row. Fig.26b reveals a wave-shaped earthquake cloud extending northeastward from mid Japan at about (37, 138.5) at 8:00 on Dec.4, 2000. By this cloud, Shou predicted to the Japanese man who had mailed the image to him that an earthquake of magnitude ≥5 would strike within 48 days in an area within ~ ±2° from the vapor source. On Jan.4, 2001, an M5.5 earthquake struck at (36.98, 138.62). This earthquake was the only one of magnitude ≥5 in an area of ±2° around the epicenter within 257 days from Jul.16, 2000 to Mar.29, 2001.

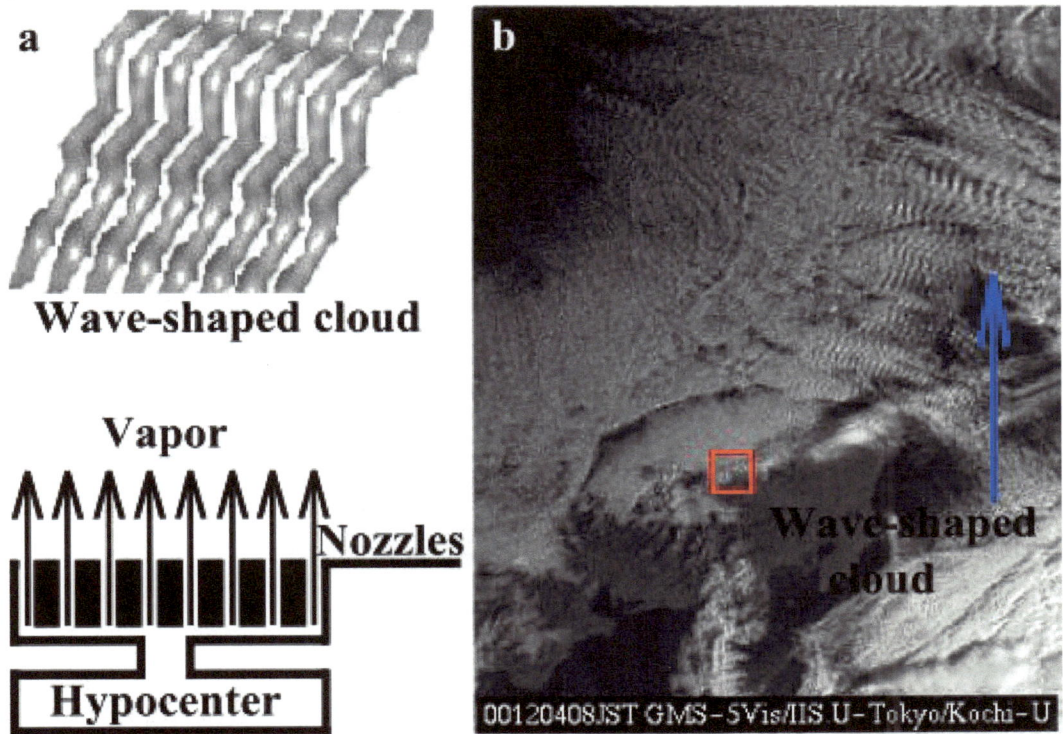

Fig.26 Wave-shaped earthquake cloud
(a) Wave-shaped earthquake cloud schematic diagram. When earthquake vapors erupt from a row of nozzles, a wind could facilitate the formation of a wave-shaped earthquake cloud.
(b) Red square plots an M5.5 epicenter (36.98, 138.62) in Japan on Jan.4, 2001. From the epicenter, a wave-shaped earthquake cloud (arrow) emerged, moved northeastward, and became bigger at 8:00 on Dec.4, 2000. The satellite image was from Kochi University@15 The earthquake data were from the USGS.
@15 http://weather.is.kochi-u.ac.jp/archive-e.html

2.2.3 Rocket, feather, lantern and spot -shaped earthquake clouds

Fig.27 presents six photographs of different appearances of earthquake clouds, taken by Shou from Pasadena, California (Harrington and Shou, 2005). Fig.27a shows the Northridge cloud which existed for only 35 minutes from 15:15 (LT 7:15) to 15:50 (LT 7:50) on Jan.8, 1994. This cloud had a large angle of elevation, which was indicative of a short distance between the epicenter and Pasadena. Shou estimated the distance between the epicenter and Pasadena to be 30~50km, but extended it to within 100km for a prediction. Shou knew that he would experience a strong earthquake.

Fig.27 Various Shapes of Earthquake Clouds
Six different shapes of earthquake clouds were photographed by Shou from Pasadena (34.14, -118.14), California. Under each photo are the date and the direction Shou took the photo.
(**a**) The line-shaped cloud on Jan.8, 1994 predicted the M6.7 Northridge earthquake (34.21, -118.53) in the same direction on Jan.17.
(**b**) The wave-shaped cloud moved from northwest to northeast on Feb.13, 1994, and predicted the M5.3 Northridge earthquake on Mar.20.
(**c**) The line-shaped cloud appeared suddenly from northwest at 2:00 on Sep.1, 1994 (LT 18:00 on Aug.31). It had a row of tails, one of which was arrowed. This cloud predicted the M7.1 earthquake (40.40, -125.68) off Coast of Northern California at 15:15 on Sep.1.
(**d**) The feather-shaped cloud from northwest to northeast predicted the M6.3 earthquake (43.51, -127.42) off Coast of Oregon on Oct.27, 1994.
(**e**) The lantern-shaped cloud from northwest predicted the M6.8 earthquake (40.55, -125.53) off Coast of Northern California on Feb. 19, 1995.

(f) The radiation-patterned cloud rising up northeastward predicted the M4.4 Joshua Tree earthquake (34.59, -116.28) in the same direction on Aug.14, 1996. All these clouds were not described by meteorology (Ahrens, 1991), but both wave-shaped and radiation-shaped clouds had been denoted earthquake clouds by Chinese and Japanese scientists in 1979 (Li, 1982).

At 12:30 (LT 4:30am) on Jan.17, 1994, Shou felt his bed rising up. Immediately, he called "Earthquake!" to wake up his daughter. Simultaneously, the land rose up and sank down about a foot each a few times, and then shocked forward and backward about a foot each for a few times. Together 30 seconds, he could not move a step. Fortunately, their apartment was OK. They and many neighbors went out of the building when a dog of their neighbor began to bark. A few days later, a man from Northridge told Shou that his bed had risen up for about 3 meters. This experience motivated Shou to perform earthquake predictions.

Fig.27b reveals a wave-shaped cloud on Feb.13, 1994. Using this cloud, on Mar.15 Shou correctly predicted the M5.3 Northridge earthquake on Mar.20 (Table 11 No.3).

Fig.27c presents a linear cloud suddenly appearing northwest of Pasadena at 2:00 on Sep.1, 1994 (LT 18:00 on Aug.31). It was a deformation of a wave-shaped cloud when a wind blew parallel to a row of erupting nozzles that formed a row of tails, one of which was pointed by arrow. Only thirteen hours later, followed the M7.1 earthquake (40.40, -125.68) off Coast of Northern California at 15:15 on Sep.1.

Fig.27d shows a feather-shaped cloud. By it, Shou correctly predicted on Oct.18, 1994 an earthquake of magnitude ≥5 in USA within 25 days (Table 11 No.9). Fig.27e showed a rare cloud consisting of a wave-shaped cloud and a lantern-shaped cloud that had a geoeruption and a line-shaped cloud inside at a higher altitude. Fig.27f depicted a radiation-pattern-shaped cloud, whose vapor might have come from many small nozzles arranging in an arc to form a vapor geyser, or a bunch of spot-shaped clouds that lasted two hours at least.

2.2.4 geoeruption-linked earthquake cloud

Fig.28 (Harrington and Shou, 2005), a photo looking towards the north of Pasadena on Aug.3, 1997 shows a cloudless line marked 4 that appeared in the midst of clouds and became a linear cloud six minutes after the photo was taken. Before the photo was taken, four cloudless lines had emerged rapidly, much faster than a jet trail. Two, marked 1 and 2, had entirely become line-shaped clouds and one, marked 3, had partially become a cloud for about three minutes.

Four earthquakes of magnitude 3.2, 4.2, 4.9 and 4.8 occurred in (38~39, -119~-118) on Aug.16~21, and they were the only earthquakes of magnitude ≥3 in this area within 310 days from Apr.21, 1997 to Feb.24, 1998. The three M4 earthquakes had high temporal and spatial coincidence, all on the same day of Aug.21 and in the same small area of (38.57±0.01, -118.49 ±0.01). The even width, straight features, the rapid emergence, the rapid transition from cloudless lines to white clouds, and the high coincidence between this atmosphere phenomenon and subsequent earthquakes strongly support the vapor theory.

Fig.28 Linear Geoeruption-Earthquake Clouds
This photo, taken by Shou from Pasadena, California toward the north on Aug.3, 1997, shows four lines that had appeared about 10, 8, 3, and less than 1 minute respectively, before Shou took the photo. They all emerged suddenly looking like Line 4, straight, even width, and cloudless in the midst of clouds. They each took about 6 minutes to become a white, linear cloud. In the photo, Lines 1 and 2 had already become clouds, and Line 3 had partially become a cloud, while Line 4 became a cloud 6 minutes after Shou took the photo. Four earthquakes of magnitude 3.2, 4.2, 4.9 and 4.8 occurred in (38~39, -119~-118) on Aug.16~21. Earthquake data are from the SCEDC

2.2.5 Multiple earthquake vapors combined into one cloud

Vapors from several spatially proximal epicenters can erupt through a common main crevice to form a linear earthquake cloud (Fig.29a). Fig.29b offers an example of a linear cloud, 590km in length, above the mid Indian Ocean at 10:00 on Dec.24, 1999. By this cloud, Shou predicted an earthquake of magnitude ≥7 in the Indian Ocean with a latitude south of -20 from Dec.27, 1999 to Feb.11, 2000. He further stated that the epicenter was more likely to be of latitude -34~-24 and longitude 60~80 (Table 11 No.41). However, not one M7, but six M5~5.7 earthquakes occurred in the time and the refine area window instead. The six earthquakes were the only sextuplets of magnitude more than or equal to 5 in the entire Indian Ocean within 3,000 days from May27, 1994 to Aug.12, 2002 (Harrington and Shou, 2005).

This phenomenon can also occur with geoeruptions. For example, two geoeruptions in Turkey on Feb.23, 2000 predicted two M4 earthquakes at Bulge B on Apr.2, 2000 and another two M4 earthquakes at Bulge C on May.7 (Fig.11, Table 2). A geoeruption at Hollister, California on Mar.20, 2001 predicted a triplet of M4 earthquakes at Hollister on Jul.2~3 (Fig.20d).

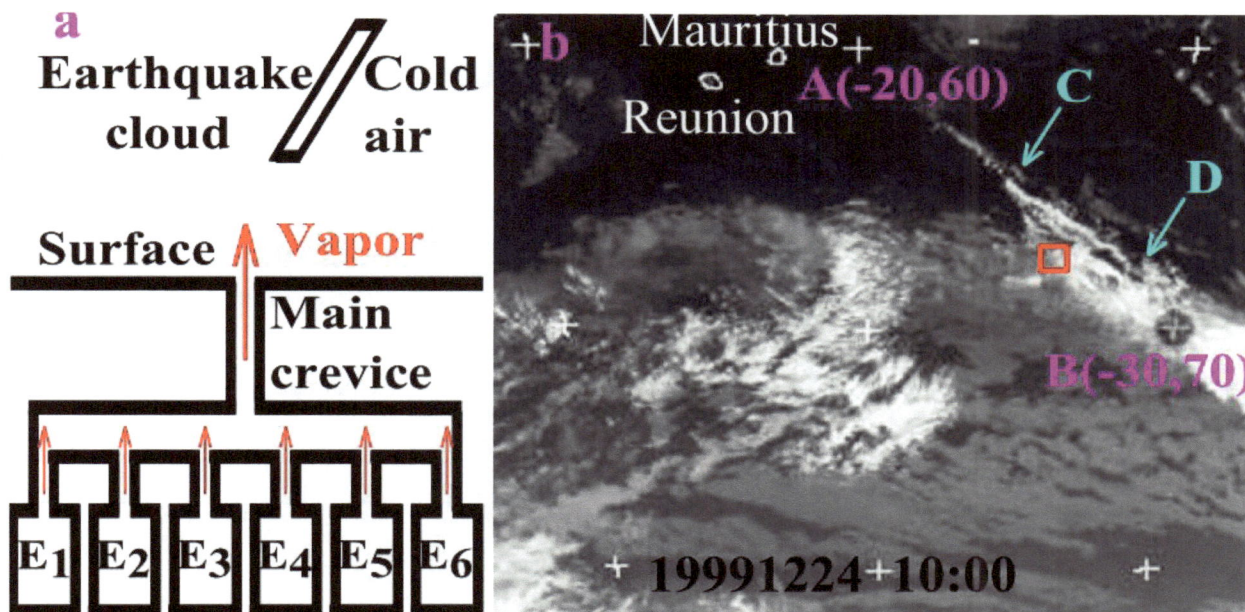

Fig.29 Combined Indian earthquake cloud
(a) Schematic diagram for independent vapor nozzles jointly forming an earthquake cloud. E_1~E_6 indicate several spatially proximal hypocenters from which earthquake vapors erupt (small red arrows) through a common main crevice (big red arrow) to the surface. The vapor then rises up to form a linear earthquake cloud after encountering cold air.
(b) Cyan arrows point at a linear cloud, 539km in length, in an area (-30~-20, 60~70) of the Indian Ocean at 10:00 on Dec.24, 1999. By this cloud, Shou predicted an M7 earthquake in the Indian Ocean south of -20° latitude from Dec.27, 1999 to Feb.11, 2000. He further predicted that the earthquake would more likely strike in a smaller refined area of (-28~-25, 60~80). Not one M7, but instead six M5~5.7 earthquakes occurred in the refined area on Feb.9~10, 2000. Red squares plot the sextuplets. Satellite image and earthquake data are from DU and USGS respectively.

2.2.6 Linear earthquake cloud

Fig.30 reveals examples of linear clouds. Common properties that distinguish these clouds from other clouds are their even width, long length, and sudden appearance. Fig.30a reveals a long, linear cloud in the west sky of Hangzhou, China at UTC 3:45~8:00 (LT 11:45~16:00) on Jun.20, 1990. It was the only cloud in the sunny blue sky. Meteorology could not explain its sudden appearance, even width, and long length. Based on this cloud, Shou predicted to two colleagues an imminent big earthquake of the magnitude of Tangshan earthquake NWW of Hangzhou. Shou wanted to photograph this special cloud, but his Chinese camera could not capture it. Seventeen hours later, an M7.7 earthquake hit Rudbar (36.96, 49.41), Iran at UTC 21:00. This earthquake was the only one of magnitude ≥7.7 in northwest (30~90, 0~120) of Hangzhou for more than 18 years from Sep.17, 1978 to May 9, 1997. This is Shou's first successful earthquake prediction (Appendix2).

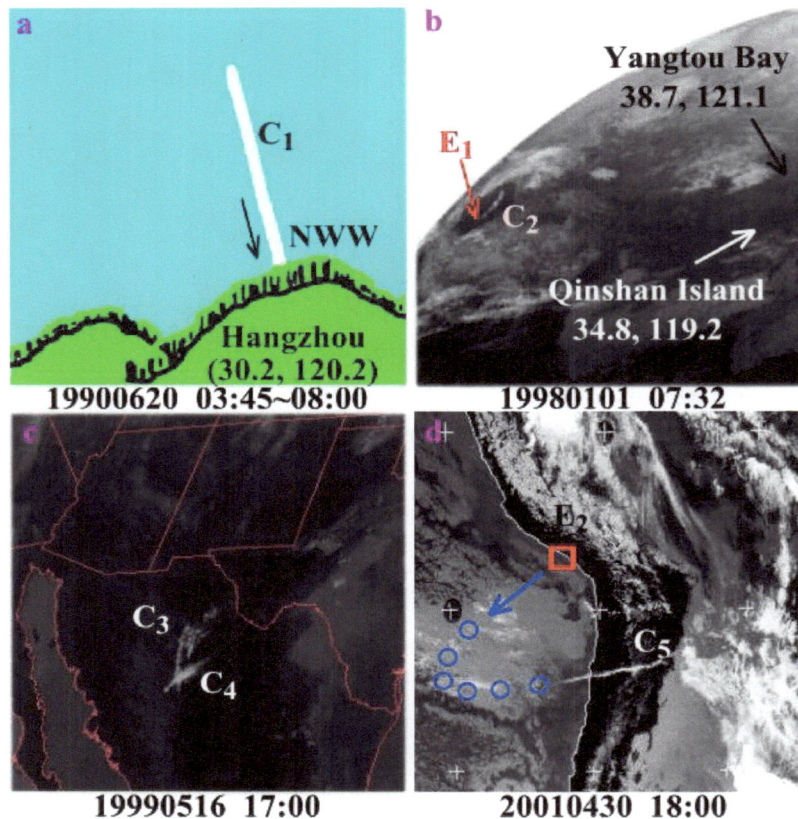

Fig.30 Linear earthquake clouds

(**a**) A long, big, tilted, linear cloud C_1 occurred in the west sky of Hangzhou (30.2, 120.2), China at UTC 3:45 on Jun.20, 1990 (LT 11:45). It was the only cloud in the sky. Its tail pointed toward NW of Hangzhou more toward West (NWW Arrow).

(**b**) A lantern-shaped cloud C_2 inside a big geoeruption amidst a vast weather cloud at 7:31 on Jan.1, 1998. Arrows point at Yangtou Bay (38.7, 121.1) and Qinshan Island (34.8, 119.2). On Feb.4, an M6.1 earthquake hit Rustaq (36.83~37.31, 69.5~70.11), Afghanistan (arrow). The GMS satellite image was from University College London (UCL@16). The earthquake data were from the UN Office for the Coordination of Humanitarian Affairs@17).

(**c**) Two linear clouds C_3 and C_4 appeared above Mexico at 17:00 on May 16, 1999. By C_3, Shou predicted an earthquake of magnitude ≥ 5 in Mexico (<20N) from May 17 to Jul. 4. Two earthquakes hit Mexico: an M7 at (18.39, -97.44) on Jun.15 and an M6.3 at (18.32, -101.53) on Jun. 21. The satellite image was from Space Science & Engineering Center of Univ. of Wisconsin-Madison (SSEC@18)

(**d**) A linear cloud C_5 appeared above Chile at 18:00 on Apr.30, 2001. Its length was 650 km. Red square E_2 plotted the M8.4 Peru earthquake (-16.26, -73.64) on Jun.23. The satellite image was from DU. Earthquake data were from the USGS except the Afghanistan quake

@16 ftp://weather.cs.ucl.ac.uk/Weather/

@17http://wwwnotes.reliefweb.int/websites/rwdomino.nsf/069fd6a1ac64ae63c125671c002f7289/60adec26e8c12cd ec12565c500395fba?OpenDocument

@18 http://www.ssec.wisc.edu/

Fig.30b presents a lantern-shaped cloud. When erupting vapor rose up, its heat dissipated part of the vast weather cloud to form a big geoeruption. The vapor continuously rose up and at higher altitude cooled down to form a linear cloud. By this analysis, Shou believed it to be an earthquake cloud. Because the satellite image did not plot coordinates, he inferred from two places: Yangtou Bay (38.7, 121.1) and Qinshan Island (34.8, 119.2) to estimate the location of the epicenter and the length of the cloud to estimate magnitude. Shou predicted to the USGS an earthquake of magnitude ≥6 in Afghanistan and its neighborhood, within a coarse window of 25~41N and 53~105E from Jan.5 to Feb.18, and a fine window of 30~37N and 58~95E from Jan.5 to Feb.4, 1998 (Table 11 No.24 Appendix4).

On Feb.4, Shou's prediction was materialized: an M6.1 earthquake hit Rustaq (36.83~37.31, 69.5~70.11) in Afghanistan (Fig.30b, red arrow). If the image had been from the IODC satellite which images Indian Ocean and offered by DU which adds grids of coordinates (Ref. Fig.7), the lantern-shaped cloud would have been in middle of the image and estimated with a much higher precision. In this case, the prediction area window could reduce 20 times at least.

Fig.30c shows two linear clouds C_3 and C_4 above Mexico at 17:00 on May 16, 1999. By C_3, Shou predicted an earthquake of magnitude ≥5 in Mexico (<20N) from May 17 to Jul.4. Two earthquakes hit Mexico: an M7 at (18.39, -97.44) on Jun.15 and an M6.3 at (18.32, -101.53) on Jun.21. Although their epicenters were 5.5° outside of the image, his prediction was correct. Shou's prediction underestimated magnitude because the image was too limited in scope to allow him to find the entire linear clouds.

Fig.30d reveals a linear cloud above Chile on Apr.30, 2001. Its length of 650 km predicted an earthquake of magnitude ≥7.5. Its tail pointed toward the Pacific Ocean. However, no big earthquake happened in the ocean, but an M8.4 earthquake hit Peru on Jun.23. A hypothesis to explain this phenomenon is that the vapor of the M8.4 Peru earthquake erupted toward the ocean and then a wind pushed it to Chile in the direction of the arrow followed by blue circles. This example implied a problem in satellite data. A satellite produced 96 images a day, but satellite owners only allowed DU to offer to the public 4~8 images a day. The low frequency of images made it very difficult to trace earthquake vapor from its origin to disappearance.

2.3 Some special examples
2.3.1 Geobulge

Fig.31 depicts an uplift of a weather cloud when it encountered a geoeruption from the Mediterranean Sea. This phenomenon happened above the Northern Atlantic Ocean near Spain on Mar.7~9, 2003. The geoeruption coincided with the M6.8 Algeria earthquake (36.88, 3.73) on May 21, 2003. It was the only one corresponding to the geoeruption in time, location and magnitude. This phenomenon is helpful for finding an earthquake, easy to observe in water vapor channel, and denoted as geothermal bulge or geobulge. A geobulge forms because a geoeruption uplifts a weather cloud.

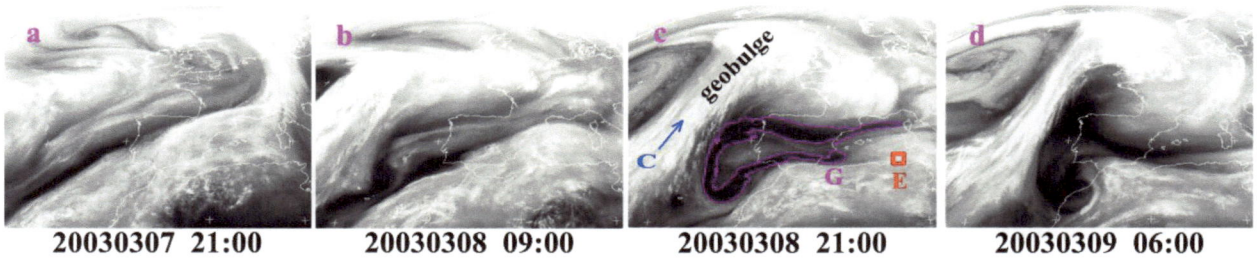

| 20030307 21:00 | 20030308 09:00 | 20030308 21:00 | 20030309 06:00 |

Fig.31 Geothermal Bulge or Geobulge
This series of images was taken in water vapor channel (5.7~7.1μm) around Algeria, Spain, and the Atlantic Ocean (30~90, -20~10) from 21:00 on Mar.7, 2003 to 6:00 on Mar.9. A geoeruption (black current) from the Mediterranean Sea passed through the Gibraltar Channel into the Atlantic Ocean and encountered a northeastward cloud. This geoeruption G expanded and gradually uplifted the cloud C to form a geothermal bulge or geobulge (Fig.31c). Red square E plots the M6.8 Algeria earthquake. The images and earthquake data were from DU and the USGS respectively

2.3.2 Earthquake cloud and geoeruption mixture

Fig.32a shows several gray, even-width curve-shaped clouds ($C_{1~5}$) above the Northeast Pacific Ocean at 21:00 on Jan.15, 2013. Meteorology cannot explain these clouds. A swarm of six moderate earthquakes near Canada (Red square E_1) on Feb.25~Mar.20 suggest these clouds being earthquake clouds.

On Aug.3, 1999, a team of ten scientists from NASA and USGS predicted the next major quake to be in Los Angeles using the method of geodetic. Immediately, many people asked Shou about the credibility of this prediction. As a response, he posted an essay "California Earthquake Situation Analysis" on Aug.10 (Appendix12) with an image taken at 9:00 on Jul.26 (Fig.32b). In the essay, Shou predicted that the next major quake would be in the black triangle '**ABC**' or the black region '**de**' at the boundary of California and Nevada. False color black indicated high temperature. Los Angeles was much cooler than the two black places and unlikely to be the epicenter for the next major earthquake. On Oct.16, an M7.4 earthquake at Hector Mine (34.6, -116.3) claimed Shou's success.

Fig.32c describes a linear cloud, 550km in length, on Sep.25, 1999. Its length predicted an earthquake of magnitude ≥7. Red square E_2, red circle T, magenta circle P, black circle B and blue circle L plot Hector Mine, Twentynine Palms, Palms Springs, Barstow and Los Angeles respectively. The same colors are used to present daily maximum temperatures in the four cities in Fig.32d that shows Los Angeles (Blue) being much cooler than the three towns near the Hector Mine.

An interesting phenomenon happened at Twentynine Palms (Red), the closest town to the epicenter. It usually recorded weather data regularly, but skipped 13 days' data from Oct.6 to Oct.18. Were these temperature data too abnormal to record? These three towns are not always warmer than Los Angeles, as shown by the temperature data on Oct.31 (Blue arrow).

20130115 21:00 **19990726 09:00** **19990925 08:00**

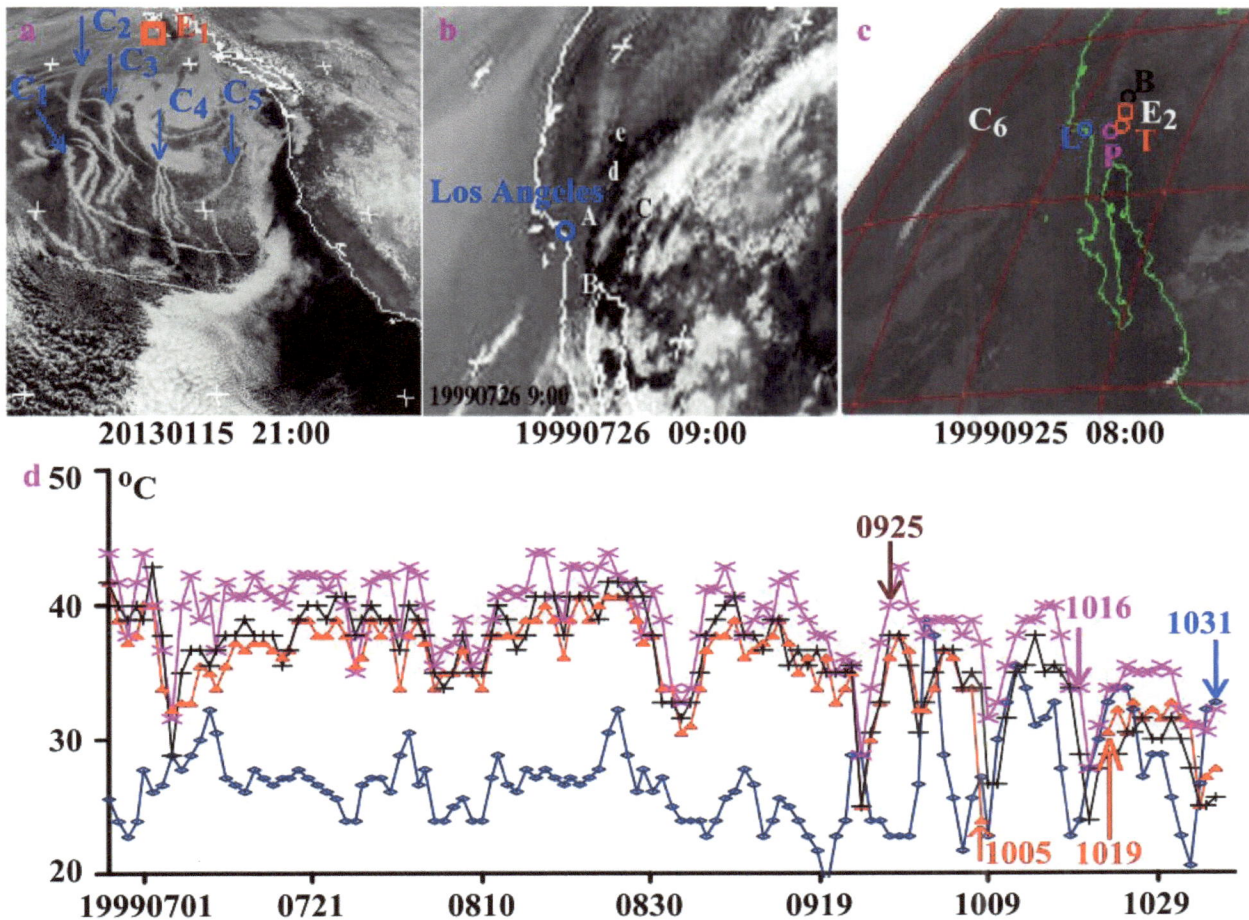

Fig.32 Special examples

(**a**) Several gray, curve-shaped clouds (C_1–C_5) appeared above the Northeast Pacific Ocean at 21:00 on Jan.15, 2013. Red square E_1 plots a swarm of six earthquakes (M4.1~4.6) near Canada from Feb.25 ~Mar.20.

(**b**) At 9:00 on Jul.26, 1999, two black areas ('**ABC**' in California and '**de**' at the boundary between California and Nevada) were much blacker and thus hotter than Los Angeles (Blue circle). On Aug.10, Shou posted this image in his website@5 to predict that the next major quake would not be in Los Angeles, but in the two hot places.

(**c**) A linear cloud C_6 near California on Sep.25, 1999, 550km in length, predicted an M7 earthquake on Oct.16, 1999. Red square E_2 plots the M7.4 Hector Mine epicenter (34.6, -116.3). Circles L (Blue), B (Black), T (Red) and P (magenta) respectively plot weather stations in Los Angeles (34, -118.2), Barstow (35.3, -116.6), Twentynine Palms (34.2, -116), and Palms Springs (33.9, -116.6).

(**d**) Curves in blue, black, red, and magenta respectively reveal daily maximum temperature in Los Angeles, Barstow, Twentynine Palms and Palms Springs from Jul.1, 1999 to Oct.31. Brown arrow (0925) and magenta arrow (1016) respectively show the appearance date of the earthquake cloud C_6 and the Hector Mine earthquake. Temperature data were skipped in Twentynine Palms between Oct.5 and Oct.19 (red arrows 1005 and 1019). Blue arrow shows temperatures on Oct.31 when Los Angeles was warmer than the three other cities. Images of Fig.32a~b and Fig.32c were from DU and the NOAA respectively. Earthquake data and temperature data were from the USGS and the WU respectively

2.3.3 Time interval between vapor eruptions and earthquakes

Fig.33a reveals a linear cloud over Xinjiang, China at 2:30 on Nov.2, 2004. Shou circled the epicenter and predicted an earthquake of magnitude ≥6 in the circle within 96 days to the public on Nov.9 or 103 days after the cloud appearance. However, no earthquake happened during the period, but an M6.2 earthquake occurred exactly at the nozzle (41.72, 79.44) arrowed by E, on Feb.14, 2005 or the 104th day after cloud appearance.

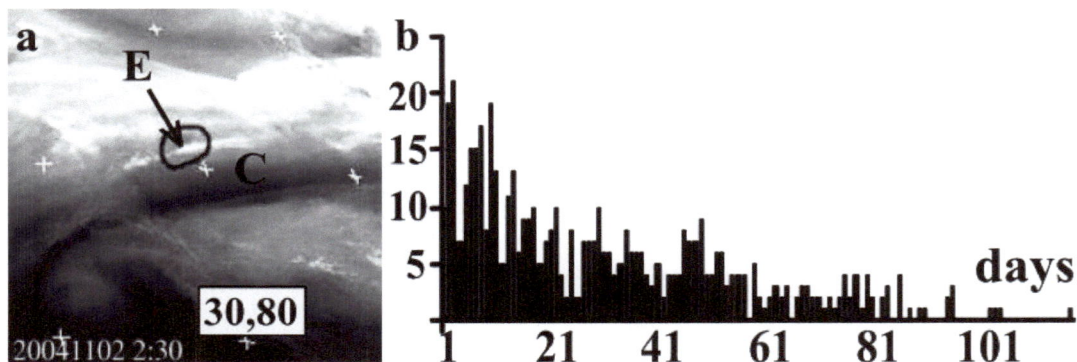

Fig.33 The distribution of durations between vapor eruptions and earthquakes
(**a**) A linear cloud C appeared in Xinjiang, China at 2:30 on Nov.2, 2004. Shou predicted to the public on Nov.9 an earthquake of magnitude ≥6 in the circle within 96 days, or 103 days after cloud appearance@5. On Feb.14, 2005, 104 days after cloud appearance, an M6.2 earthquake occurred at Aksu (41.72, 79.44) arrowed by E, exactly in the predicted area.
(**b**) A histogram of durations between vapor eruptions and earthquakes. In a total of 509 data, the average is 30 days, and the longest is 118 days. In the 509 earthquakes, 50 (about 10%) happened within the first three days. All above data were from Earthquake Cloud & Short Term Prediction@5

This earthquake has been the only one of magnitude ≥6 in the predicted area since Dec.2, 2003. Based on historical earthquake data of this area, random predictions with the same predicted area and magnitude and a 96-day time window will have a probability of 5.3% or 1 in 18.8 to be successful.

During the 104 days from the cloud appearance to the earthquake, this earthquake was the only one of magnitude ≥6 in the region of -90~90 and 30~90, or 1/6 of the Earth. This region contains Turkey, Iran, Caucasus, the Black Sea, the Caspian Sea, Turkmenistan, Pakistan, Afghanistan, Kazakhstan, Tajikistan, Kyrgyzstan, Mongolia, Western China, India, Yemen, Oman, Tanzania, South Africa, the Indian Ocean, and so on, and is a very active region for earthquakes. A random area guess would have little chance to be successful.

Thus, the M6.2 earthquake was a consequence of the same process that generated the cloud. Mistakes due to for example a slightly too small time window do not indicate a problem with the precursor, but are a fundamental part of developing a method empirically (Shou, 2006a).

The duration from vapor eruption (an earthquake cloud or a geoeruption) to its subsequent earthquake is unknown. Shou spent 15 years from 1990 to 2004 trying to find the longest delay. Fig.33a gives an important example to reveal how Shou underwent the difficulty.

Fig.33b reveals a histogram of durations relying on 509 data, whose average is 30 days and the longest is 118 days. There is only one earthquake with an 118-day duration. Because the USGS has data losses (Table 5), it is possible that for this incidence, an earthquake corresponding to the vapor eruption occurred sooner than 118 days. We thus select 112 days (the average of 118 days and 106 days) as the possible longest duration (Shou, 2006b). Afterward, thousands durations were checked, and none surpassed 112 days. However, it is difficult to prove 112 days as the longest duration.

In Fig.33b, 10% of them happened in the first three days. Some earthquakes happened very quickly. For example, the M7.7 Rasht, Iran earthquake killed 50,000 people and injured 320,000 only 17 hours after the appearance of the cloud (Fig.30a). For evacuation, it is important to have high resolution and frequent satellite images.

2.3.4 Magnitude calibration

Fig.34 reveals an earthquake cloud lasting for 10 hours in Iran on May 29, 2003, which preceded an M5.6 earthquake. Comparing the Dec.20~21, 2003 Bam cloud which lasted 26 hours with this cloud, Shou decided to increase the magnitude of his Bam prediction made on Dec.25 from '≥5.5' to '≥6.5'. But before he could modify the prediction and alert Iranian people, two messages had already arrived at his email account, congratulating him on the accuracy of his prediction. This and many other examples illustrate three important points for the accuracy of predictions.

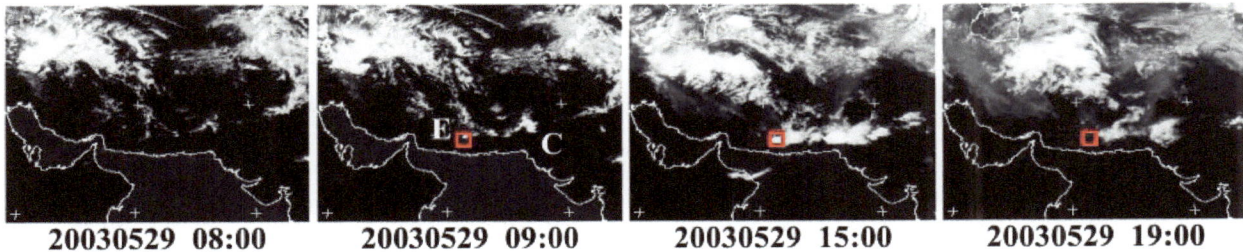

| 20030529 08:00 | 20030529 09:00 | 20030529 15:00 | 20030529 19:00 |

Fig.34 Reference of magnitude
Red square plots the epicenter of an M5.6 earthquake at (27.3, 61), Southern Iran on Jun.24, 2003. An earthquake cloud C emerged from this impending epicenter at 9:00 and lasted until ~19:00 on May 29, 2003. The images and earthquake data were from DU and the USGS respectively.

First, the ability to compare is important. Thus, for vapors observed for the first time (e.g. Fig.5, 6, 11, 13, 18, 20~24, 27~32, etc), the lack of previous examples to compare to made prediction difficult. With enough examples, empirical coefficients may be derived which can be very useful. For linear clouds, (Shou, 2006a) reported that the length of 300km and 350km predicted a magnitude 6 and 7 respectively. For cases with continuous vapor eruption which don't form linear clouds, Shou estimates that 10 hours and 24 hours of vapor eruption will predict a magnitude 5 and 6 respectively. The above coefficients were derived from low-frequency satellite images and non-standardized magnitude data, so they have errors.

Second, predictions must be made quickly. For example, the M7.7 Iran earthquake struck only 17 hours after its cloud (Fig.30a); the M7.1 California earthquake occurred only 13 hours after its cloud (Fig.27c), etc.

Third, accurately measured earthquake data are important for establishing a correct correlation between earthquakes and predictions and for fairly judging earthquake predictions. All measured earthquake data have two kinds of errors: systematic and random. Table 6 (Shou, 2006a) shows Rank "A" (smallest error) measurements made by 17 organizations on the magnitudes of the same Indonesia Tsunami earthquake on Dec.26, 2004. The range was enormous, ranging from M5.5 to M8.7.

The only way to solve this problem is to calibrate all seismometers with an independent standard. For example, Shou (2006a) suggested the creation of an artificial standard, e.g. a chemical explosion whose energy, latitude, longitude, and depth are specifically designed for this purpose. So far, this problem still persists. One may argue that data from the USGS are precise. It is impossible to obtain accurate data without a standard. For example, the USGS gave the earthquake that triggered Indonesia tsunami a magnitude 8 before the tsunami (Marris, 2005), but 9 afterward.

2.3.5 Earthquake data problems

Besides inaccuracy in earthquake measurement, Table 5 and Table 7 show an additional problem of missing earthquake data. Table 5 shows 8 earthquakes, but the USGS did not report 4 (50%) of them and the CWBT did not report 2 (25%) of them. Geophysical Research Letters (GRL) offered the public two earthquake catalogs. One was from the Greek Seismological Institute-National Observatory of Athens (SI-NOA), and the other was from NOAA (PDE) for earthquakes of M≥5.0 in (35~42, 17~27) and 1987~1989. In a total of 58 data, the SI-NOA lost 12 (20.7%), and the NOAA lost 19 (32.8%). If both catalogs failed to report an earthquake, a correct prediction would be regarded as an incorrect prediction.

2.4 How to predict earthquakes

We have discussed various examples to explain how to use the earthquake vapor model to detect erupting vapor and to predict earthquakes. The general method is as following.

2.4.1 How to predict epicenter

First, select a series of continuous images from a geostationary satellite taken from the same wavelength channel, and examine whether or not vapor erupted suddenly from a fixed point, or against a normal wind. For example, we select mid-infrared (10.5~12.5μm) images of satellite IODC (0, 63) on Dec.20~21, 2003, open with the computer program "Windows Picture and Fax Viewer", put the arrow of a mouse at Bam, and continuously push the key '→'. Then, we will see an animation of a cloud from a fixed vapor source at Bam for 26 hours@6

Second, a vapor has a smaller velocity at the beginning and end of an eruption. Because of the smaller velocity, they stay near the epicenter for longer. Thus, these times are the best for finding the epicenter. For example, the M6.5 Kerman epicenter can be found at 8:00 and 16:00 on Dec.14, 2004 (Fig.8); the M6.6 Andaman epicenter can be found at 3:00 on Nov.15, 2004; the M7.5 Nicobar epicenter can be found at 3:00 on Nov.16; the M9.0 Sumatra epicenter can be found at 9:00 on Nov.16 (Fig.13). To narrow an area window, it is important to have high-resolution and frequent images.

The above method is sometimes but not always useful. For example, the Izmit epicenter (Fig.17) was not visible in the satellite images. Shou (2011) attributed this problem to an artificial limit imposed on the highest realistic atmosphere temperature, but did not find evidence until 2013. Fig.35 shows such a set limitation to $342^{\circ}K$ or $69^{\circ}C$. Thus, temperature higher than that was set to white ($0^{\circ}K$). Thus, if the limit is removed and all temperature values are plotted, then we can pinpoint the hottest place or a nozzle in warm or even hot surrounding.

2.4.2 How to predict time

We have already discussed that the longest duration from vapor eruption to its subsequent earthquake is about 112 days, with an average of 30 days (Fig.33b). We have also discussed that 10% of quakes happen within 3 days after an eruption, and other 90% within 3 days after the second eruption or the second temperature peak (Shou, 2011). Therefore, by isolating a nozzle to monitor the second eruption or the second temperature peak, we should be able to narrow the time window into a week.

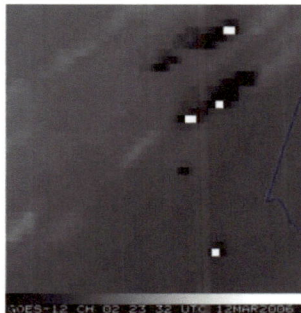

Figure 1: Example of rollover in observations by GOES-12 Imager Channel 2. The image contains an area of fires with brightness temperatures sometimes exceeding 342K. The higher the GVAR count value, the blacker the pixel. However, the observations in the hottest regions exceed the GVAR maximum and have rolled over to low values, producing the white pixels. The irregular blue line on the right is a portion of a state border superimposed on the image. (Figure courtesy of Scott Bachmeier, Space Science and Engineering Center, University of Wisconsin, Madison, WI)

Fig.35 The assumed highest atmospheric temperature
This infrared image is from NOAA at UTC 23:32 on Mar.12, 2006. Its highest temperature is artificially defined as $342^{\circ}K$ or $69^{\circ}C$, more than which has no definition. Black indicates fire area; while white indicates low temperature. The pure white areas in blacks imply temperatures surpassing $69^{\circ}C$ to show brightness of $0^{\circ}K$ because of no definition@12.

2.4.3 How to predict magnitude

The quantities of vapors may correlate with the magnitude of earthquakes. From this correlation, we may predict magnitude from the size or duration of vapor. We have just discussed that Shou (2006a) reported that the lengths of 300km and 350km of a linear cloud predicted magnitudes 6 and 7 respectively. For cases with continuous vapor eruption, Shou estimate that 10 hours and 24 hours of vapor eruption may predict magnitudes 5 and 6 respectively. We cannot evaluate magnitude precisely because of no standardized magnitude (Table 6, Shou, 2006a). Furthermore, error in reported earthquake magnitudes is large. Marris (2005) reported that the USGS gave the 2004 Sumatra earthquake a magnitude of 8 before the tsunami, but 9 after the tsunami (note that earthquake magnitude is independent of tsunami). Magnitude errors affect predictions in two ways. First, prediction of magnitudes is based on empirical correlations conducted on earlier earthquakes whose magnitudes suffer large uncertainty. Second, magnitude errors also exist in predicted earthquakes, making evaluation of earthquake predictions difficult. If we standardize earthquake data and improve their precisions, a successful evacuation will become possible.

2.5 Problems caused by satellite images

We have discussed that artificially setting the highest atmosphere temperature to 69°C causes great difficulty in distinguishing a nozzle from its hot vicinity (Fig.17a~d). Fig.36 shows another problem. Two images were from the same channel of the same satellite at the same time, but produced by different agencies. The top was from the European Organisation for the Exploitation of Meteorological Satellites (EUMETSAT) @19; while the bottom was from Dundee Satellite Receiving Station (DU) @4. Obviously, the bottom had much higher resolution. On the other hand, the top had a frequency of about 24 images a day, while the bottom only 4~8 per day because of a license limitation. As a result, the Public could not obtain high frequency and high resolution images although a satellite produced 96-high resolution images a day. Dundee also offered other global images, such as GOES West (0, -135), GOES East (0, -75), Meteosat SEVIRI (0, 0), and MTSAT (0, 145), but the same problems existed in all images.

The Northridge cloud existed for only 35 minutes form LT 7:15am to 7:50am on Jan.8, 1994 (Harrington and Shou, 2005). This kind of clouds will need half-hourly imagery to capture.

Since many large earthquakes happened shortly after vapor eruption, it is important to offer imagery in a timely fashion. The M8 Sichuan earthquake struck at 6:28 on May 12, 2008. Earthquake cloud appeared at 18:00 on May 11. However, DU issued the image at 5:12 on May 14, which was too late.

Historical abnormal surface temperature and earthquake data are important for understanding past earthquakes and predicting future earthquakes. For instance, a pulse temperature of 300°C happened in the Las Americas (18.4, -69.6), Dominican Republic at LT 13:41 on Jun.2, 1997 in the presence of a north wind. This abnormal temperature coincided with an M4.3 earthquake (18.88, -69.6) exactly north of Las Americas on Jun.8. It was the only earthquake north of Las Americas and within ±2.5° from longitude -69.6 within 148 days from Jun.2 to Oct.27. Abnormal surface temperatures in a large area may predict a large earthquake. Abnormal temperatures are important, but

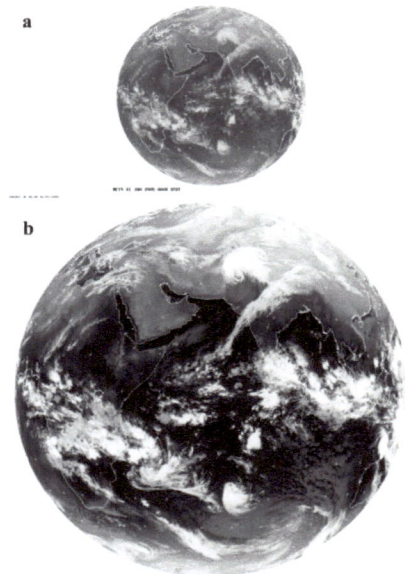

Fig.36 A comparison of the same image between EUMETSAT and DU
(a) EUMETSAT@19 owned geostationary satellite Meteosat VISSR or IODC at (0, 63). It offered the public an infrared image (Channel 2) of file size 60kb from this satellite at 0:00 on Jan.1, 2005.
(b) DU@4 offered the public the same infrared image of file size 207kb from the same satellite at the same time. The latter had a much higher quality than the former.
@19http://www.eumetsat.int/website/home/index.html

skips in temperature data seriously limit their use (e.g. Fig.15c, 17e, 18a~b, 23f, 32d etc.)

There are other satellite data problems such as missing areas in images, disagreement in whether clouds are present or not between images taken from the same satellite, channel, time and area, but offered by different sources. However, we will not discuss all of them.

If we could improve the current state, for example, setting the temperature upper-bound to about $1520^{\circ}C$ instead of the artificial limitation of $69^{\circ}C$, we would be able to pinpoint at a nozzle of erupting vapor or an impending epicenter in hot or cold surrounding environment. If we could further standardize earthquake data by setting up an artificial standard such as a chemical mine whose energy, latitude, longitude, and depth can be designed precisely, we would be able to have reliable earthquake data and predict magnitude of an earthquake more precisely. If we could further monitor the second vapor eruption or second temperature increase by isolating a nozzle, we would be able to narrow the time window of a prediction into about a week (Shou, 2011, "Summary of earthquake vapors" **1.9**). Therefore, an evacuation would be possible.

2.6 False warnings and misses

False warnings and misses are important criteria for judging how useful a precursor is. A "false warning" is the appearance of a precursor without a subsequent earthquake. For example, a team of NASA and USGS used the most modern technology to find geodetic abnormality in Los Angeles, and predicted the "next major quake" in Los Angeles on Aug.3, 1999. However, no major quake followed so far. Thus, this geodetic prediction is a false warning. A "miss" means an earthquake without a precursor. For instance, the Tangshan earthquake does not have a foreshock precursor (Zhang, 1981), so foreshock has a miss. Geller et al., (1997) published, "Earthquakes cannot be predicted", which implied that none of widely studied precursors occurred reliably before earthquakes.

Just like we test the quality of a big batch of products by testing a few samples, we select a few representative earthquakes to test whether they are linked to earthquake vapor. Northridge, which is near Los Angeles, is under the best earthquake surveillance system. This particular area had no earthquake history, and no precursors were observed for the earthquake. Therefore, the Northridge earthquake was regarded as an unpredictable example (Heaton and Wald, 1994). However, the Northridge cloud (Fig.5a) appeared like a launching rocket from Northridge. Moreover, it had abnormal temperature in downwind direction (Fig.5b~d). Thus, the Northridge cloud was non-meteorological. The high coincidence between the cloud and the earthquake rejected the dogma that earthquakes are unpredictable.

The Bam cloud and the Bam prediction offered another excellent example. The Bam cloud (Fig.6) looked like smoke from a chimney and persisted for 26 hours at Bam. Thus, it was non-meteorological. By this cloud, Shou predicted the Bam earthquake exactly (Appendix1). This quake was the first big one there in history.

We will not repeat other excellent examples discussed previously. Instead, we will discuss below whether or not all devastating earthquakes of death toll of more than 10,000 from Jan.1, 1990 to Jan.1, 2016 had the vapor precursor.

2.7 All large devastating earthquakes are preceded by vapor

Table 8 reveals nine devastating earthquakes since 1990. We have already shown in Fig.30a, Fig.17, Fig.6, and Fig.13 the vapor of the Rasht, Iran earthquake, the Izmit, Turkey earthquake, the Bam, Iran earthquake, and the Sumatra, Indonesia earthquake respectively. They were all preceded and predictable by vapors (Shou, 1999, Harrington and Shou, 2005, Shou, 2006b). We will discuss the five remaining large earthquakes.

2.7.1 The M8 Gujarat, India earthquake

For the Gujarat earthquake, Fig.37a~e shows the sudden appearance of a gray earthquake cloud C_1 from Gujarat at 18:00 on Nov.7, 2000. It moved northeastward and caused earthquake clouds C_2 and C_3 at 12:00 on Nov.8. C_3 moved eastward to form a linear cloud, 830km in length, at 9:00 on Nov.9 (Fig.37e). This length predicted a magnitude of about 8.

During each day of the eruption from Nov.7 to Nov.10, daily maximum temperature in New Delhi reached the highest record of that date in 18 years from 1996 to 2013 (Fig.37f). Similarly, the daily maximum temperature of the nearby Lahore, Lucknow, Varanasi, and Kolkata reached its respective highest in 18 years for two or three of the four days.

In Ahmedabad (23.1, 72.6), temperature records had many skips, but it increased 2°C from 33°C on Nov.5 to 35°C on Nov.7 and kept 35°C for four days (Red in Fig.37g). The temperature of 35°C was the highest in three months from Nov.1, 2000 to Jan.31, 2001. Mumbai was like Ahmedabad. The above temperatures are all denoted abnormal (Shou, 2011). The places with abnormal temperatures are plotted in Fig.37a and are downwind from the Gujarat earthquake cloud. Like Twentynine Palms (Fig.32d), Ahmedabad, the closest station to the Gujarat epicenter, skipped many temperature records before and after the Gujarat earthquake (magenta Fig.37g).

An interesting record was a pulse increase of 8°C in 40 minutes from 23°C at 10:50 (LT 16:50) to 31°C at 11:30 (LT 17:30) in Kolkata on Jan.26, 2001 (Fig.37h), when the Gujarat earthquake had happened 8 hours prior. This 8-hour interval could be caused by the time it takes for heat to be transmitted from the epicenter to the temperature measurement station. The distance between Gujarat and Kolkata was 1860 km.

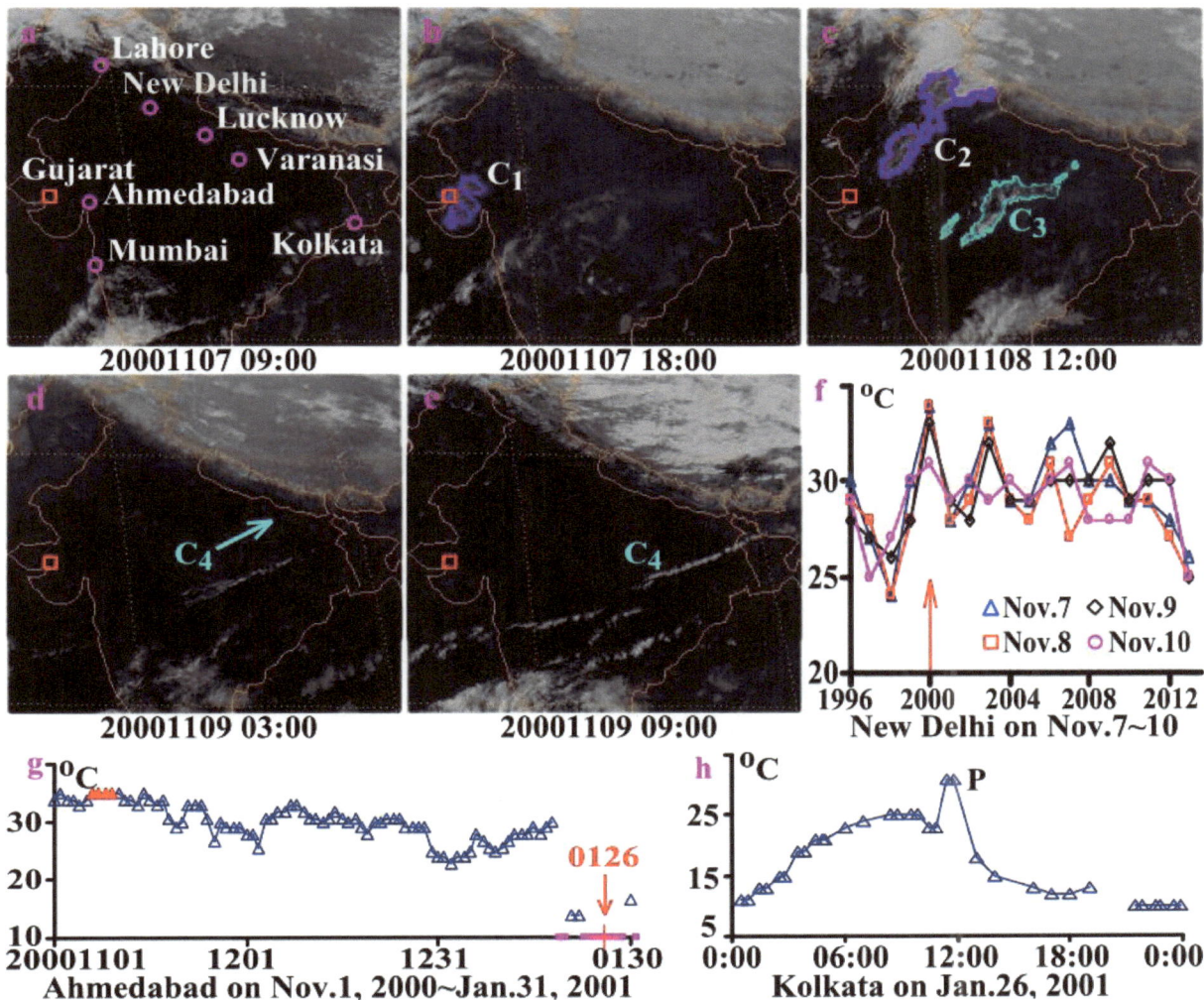

Fig.37 The Gujarat earthquake vapor

(**a**) Red square plots the M8 Gujarat epicenter (23.4, 70.2) on Jan.26, 2001. Magenta circles plot airports at New Delhi, Lucknow, Varanasi, Kolkata, Lahore, Ahmedabad, and Mumbai where abnormal temperature appeared during the Gujarat vapor eruption on Nov.7~10, 2001.

(**b**) A gray earthquake cloud C_1 (Purple) appeared near the epicenter at 18:00 on Nov.7, 2001.

(**c**) The vapor moved northeastward and caused a gray cloud C_2 (Purple) at 12:00 on Nov.8. Simultaneously, part of the vapor moved eastward to form a white cloud C_3 (Cyan).

(**d-e**) C_3 rose up, condensed into C_4 and moved eastward at 3:00 on Nov.9. It became linear of about 830km in length at 9:00.

(**f**) Blue, red, black and magenta curves reveal daily maximum temperature in New Delhi on Nov.7, 8, 9 and 10 from 1996 to 2013 respectively. Red arrow points at 2000 for the highest temperatures of all curves.

(**g**) Although many temperature records in Ahmedabad on Nov.7~10 were absent, they (Red triangles) were still the highest daily maximums in three months. Like temperature data skips in Twentynine Palms before and after the Hector Mine earthquake (Fig.32d), temperature data in Ahmedabad skipped before and after the Gujarat earthquake (Magenta). The temperature data were from the WU. The satellite images were from the National Climatic Data Center (NCDC) @20

(**h**) P shows a pulse temperature increase in Kolkata from 23°C at 10:50 to 31°C at 11:30 on Jan.26, 2001 about 10 hours after the earthquake.

@20 http://www.ncdc.noaa.gov/gibbs/year

- 55 -

2.7.2 The M7.6 Kashmir, Pakistan earthquake

Fig.38 reveals a sudden linear earthquake cloud above Xinjiang, China eastward at 9:00 on Sep.28, 2005. By this cloud, Shou predicted an earthquake of magnitude ≥5.5 in the circled area (Fig.38f) within 92 days to the public @5 at 2:25 on Oct.7, 2005 (Appendix11), An M7.6 earthquake proclaimed Shou's success at 3:50 on Oct.8.

This cloud had a length of about 600km that predicted an earthquake of magnitude ≥7.5. Afterward, Shou (2006a) published this empirical experience that a linear cloud of 300km and 350km in length predicted magnitude of 6 and 7 respectively.

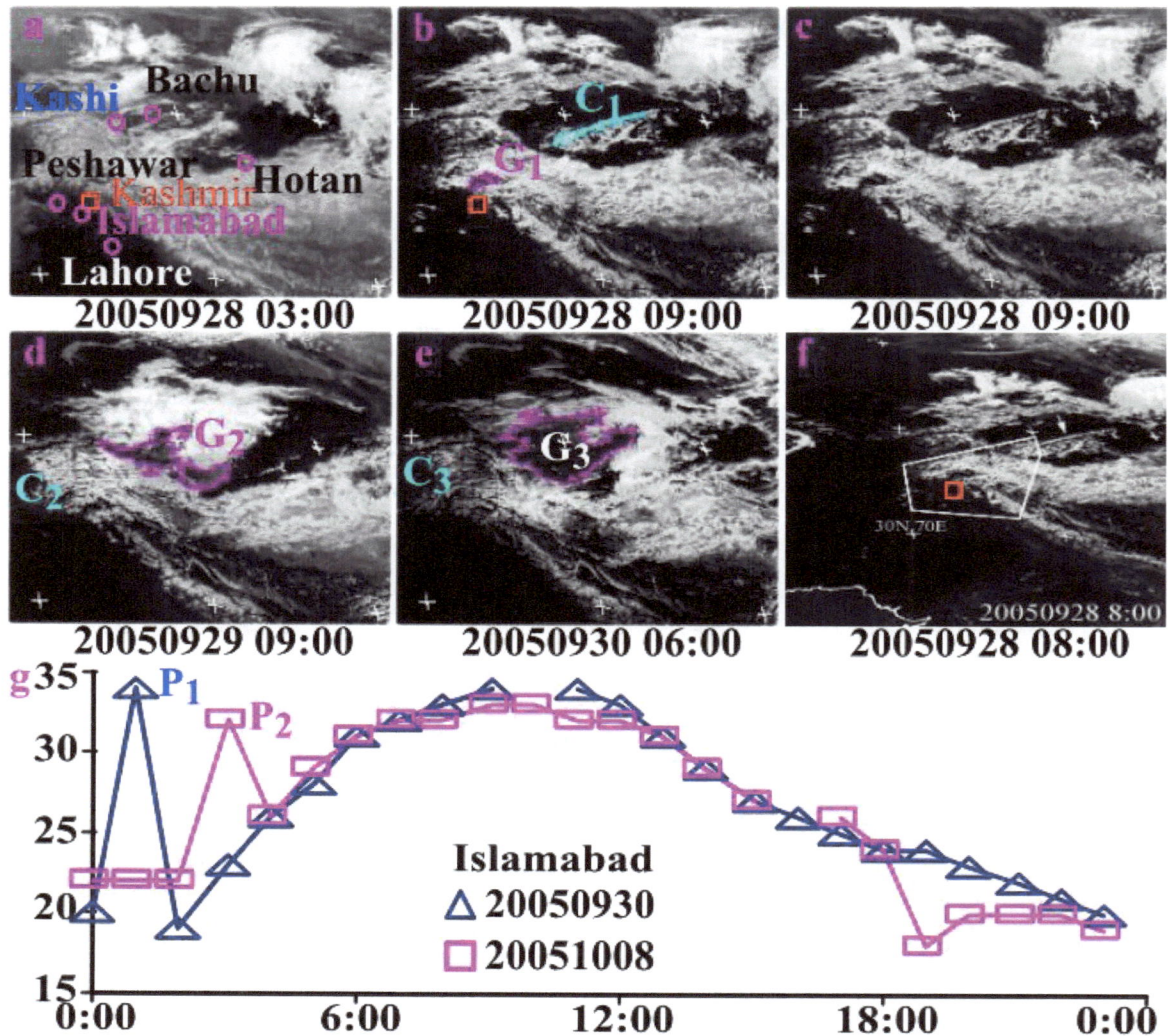

Fig.38 The Kashmir earthquake vapor

(**a**) Red square plots the M7.6 Kashmir, Pakistan epicenter (34.43, 73.58) on Oct.8, 2005. Magenta circles plot airports at Islamabad, Lahore, Peshawar, Kashi, Hotan, and Bachu. The six airports coved an area of 340,000km^2 where the average of daily maximum temperatures increased about 2°C from Sep.27, 2005 to Sep.28.

(**b-c**) A linear earthquake cloud C_1 (Cyan) appeared above Xinjiang, China at 9:00 on Sep.28, 2005. It had a length of about 600km. A geoeruption G_1 (magenta) was near the epicenter (Red square). Panel **c** reveals the unmarked original of Panel **b**.

(**d-e**) C_2 and C_3 were wave-shaped earthquake clouds, and G_2 and G_3 (magenta) were geoeruptions developed over the next two days after the initial eruption.

(**f**) Shou predicted to the public an earthquake of magnitude ≥ 5.5 in the circled area of image 200509280800xWChinaCL.jpg within 92 days in website "Earthquake Clouds & Short Term Predictions" @5 at 2:25 on Oct.7, 2005 or LT 18:25 Oct.6 (Appendix11). Red square plots the epicenter, which was exactly in the predicted area.

(**g**) In Islamabad, temperature increased $14^{\circ}C$ in one hour (P_1 Blue) from $20^{\circ}C$ at 0:00 on Sep.30 (LT 5:00) to $34^{\circ}C$ at 1:00 (LT 6:00) when vapor was erupting. Another pulse temperature increase P_2 (Magenta) lasted from $22^{\circ}C$ at 2:00 on Oct.8, 2005 to $32^{\circ}C$ at 3:00, just 50 minutes before the earthquake.

Recent investigations show that the average of daily maximum temperature increased about $2^{\circ}C$ in an area of about 340,000km^2 among Islamabad, Lahore, Peshawar, Kashi, Hotan, and Bachu (Fig.38a) from Sep.27, 2005 to Sep.28. After the linear cloud, wave-shaped clouds and geoeruptions appeared on Sep.29 and 30 (Fig.38d~e).

Consistent with these vapors, a pulse temperature increase of $14^{\circ}C$ (P_1 Blue in Fig.38g) emerged in Islamabad from $20^{\circ}C$ at 0:00 on Sep.30 (LT 5:00) to $34^{\circ}C$ at 1:00 (LT 6:00). A pulse temperature increase P_2 (Magenta Fig.38g) happened in Islamabad, rising from $22^{\circ}C$ at 2:00 on Oct.8, 2005 to $32^{\circ}C$ at 3:00, just 50 minutes before the earthquake.

2.7.3 The M8 Sichuan, China earthquake

Fig.39 shows a swarm of earthquake clouds erupting from Sichuan and Yunnan at 9:00 on May 11, 2008 (Fig.39b). It also shows two linear earthquake clouds of 300km and 740km at 18:00 of May 11 (Fig.39d), which predicted an M6.1 and an M8 earthquake on May 12.

Fig.39e~f show two linear clouds: one at 3:00 on May 17, and the other at 15:00 on May 19. From the two clouds, Shou predicted to the public on May 20, during serious cancer, two earthquakes of magnitude ≥ 6 with 110 days, one in each grey rectangle ("Earthquake Clouds & Short Term Predictions" @5). Indeed, two M6 earthquakes hit the squares: one on May 25 and the other on Aug.5.

Four strong clouds predicted four big earthquakes without any false warming. The four earthquakes were the only big earthquakes in an area of 4.2 million km^2 (latitude 20~40 and longitude 100~120) within five years and ten months from Jun.3, 2007 to Apr.19, 2013. All four big earthquakes had vapor eruptions without miss.

Fig.39g depicts daily maximum temperature changes from 119 stations in an area of 2.1 million km^2 (latitude 24~34 and longitude 100~120) from May 10, 2008 to May 11. One hundred and seventeen stations experienced temperature increases. The largest increase was $9.5^{\circ}C$ at Jiulong. The highest decrease was $-1.6^{\circ}C$ at Pingtan. The average increase was $3.4^{\circ}C$. This large increase was not only from vapor eruptions of the M8 and the M6.1 Sichuan earthquakes, but also from many moderate earthquake clouds (M4~5.9) along Sichuan and Yunnan (Fig.39b).

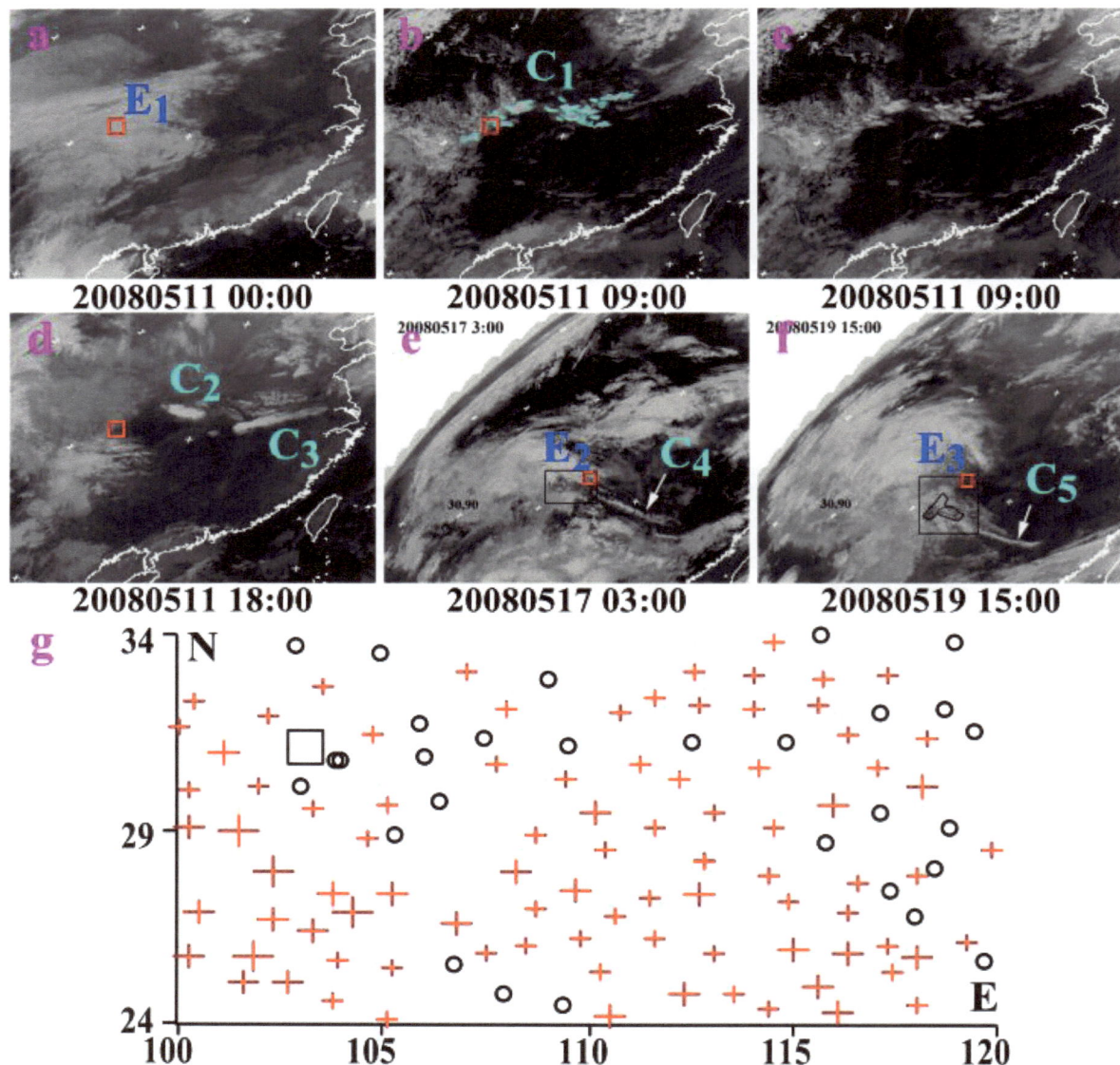

Fig.39 The Sichuan earthquake vapor

(**a**) Red square E_1 plots the epicenter of the M8 Sichuan, China earthquake (31.0, 103.3) and of an M6.1 earthquake nearby on May 12, 2008.

(**b-c**) Earthquake cloud C_1 (Cyan) appeared at 9:00 on May 11. Panel **c** shows an unmarked image of Fig.39b.

(**d**) Two linear earthquake clouds C_2 (300km in length) and C_3 (740km) appeared at 18:00 and predicted earthquakes of magnitude ≥6 and ≥7.5 respectively.

(**e-f**) Requested by many Chinese people, Shou predicted on May 20 to the public two earthquakes of magnitude ≥6 within 110 days from two linear clouds C_4 at 3:00 on May 17 and C_5 at 15:00 on May 19 respectively, with predicted areas outlined by grey rectangles ("Earthquake Clouds & Short Term Predictions" @5). Red square E_2 plots the epicenter of an M6 earthquake (32.7, 105.5) on Aug.5. Red square E_3 plots the epicenter of an M6 earthquake (32.6, 105.4) on May 25.

(**g**) Black square plots the M8 Sichuan epicenter. Black circles, and red small, medium, and large crosses plot stations of daily maximum temperature change in the ranges of -1.9~1.9, 2~4.9, 5~7.9 and 8~10.9 from May 10 to May 11 respectively. In 119 stations, 117 experienced an increased temperature. The highest increase is +9.5°C at Jiulong (29, 101.5). The highest decrease is -1.6°C at Pingtan (25.5, 119.8). The average is 3.4°C. The satellite data, earthquake data and temperature data were respectively from DU, the USGS and the NCDC.

2.7.4 The M7.3 Haiti earthquake

Fig.40a describes a geoeruption G, and an earthquake cloud C_1 near Haiti at 12:00 on Dec.29, 2009. Fig.40b shows another earthquake cloud C_2 at 3:00 on Dec.30, moved southwestward at 6:00 (Fig.40c) and formed C_3 at 9:00 (Fig.40d). This linear, loose cloud had a length of about 650km and predicted a magnitude more than 7.

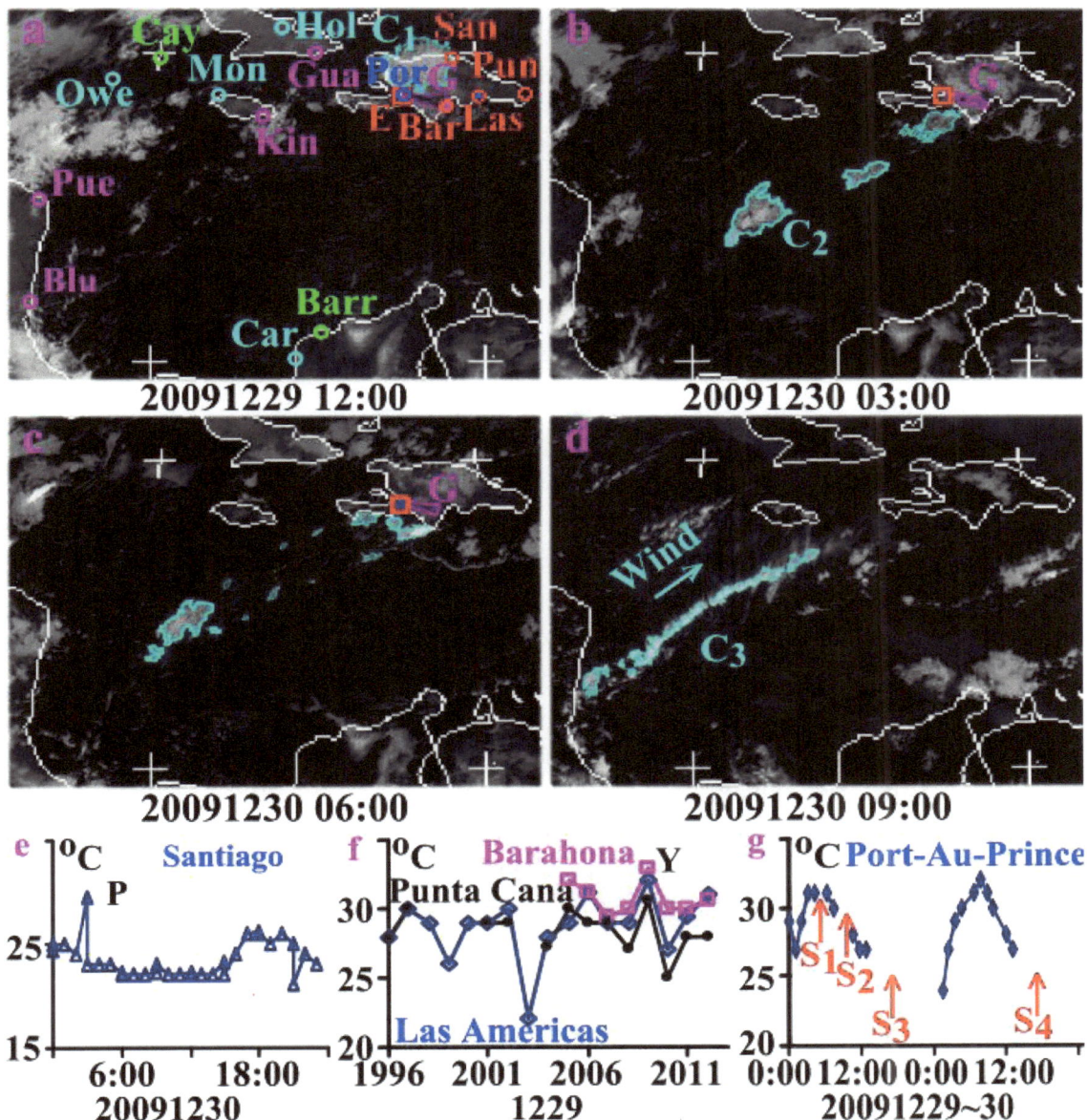

Fig.40 The M7.3 Haiti earthquake vapor
(**a**) Red square E plots the M7.3 Haiti earthquake (18.44, -72.54) on Jan.12, 2010. Geoeruption G (Magenta) and earthquake cloud C_1 (Cyan) appeared near E at 12:00 on Dec.29, 2009. G was maintained in Fig.40b~c. Red, magenta, blue, green, and cyan circles respectively plot airports with temperature abnormality ("*Definition of Abnormal Temperature*" 1.8; Shou, 2011), temperature increases of at least 1°C with relatively complete records, temperature increases of at least 1°C with incomplete records, no temperature changes, and temperature decreases from Dec.29, 2009 to Dec.30. Bar, Las, Pun, San, Gua, Kin, Pue, Blu, Por, Cay, Barr, Car, Hol, Mon, and Owe

- 59 -

respectively plot Bara hona, Las Americas, Punta Cana, Santiago, Guantanamo Bay NAS, Kingston, Puerto Limon, Blue fields, Port-Au-Prince, Cayman Brac, Barranquilla, Cartagena, Holguin, Montego Bay, and Owen Roberts.

(**b-c**) The vapor became strong, and erupted southwestward to form cloud C_2 at 3:00~6:00 on Dec.30.

(**d**) C_2 moved westward, and a west wind (Cyan arrow) blew it northeastward to form cloud C_3. The length of C_3 reached that for magnitude 7.

(**e**) P shows a pulse temperature increase from 24°C at 2:00 on Dec.30 (LT 22:00 on Dec.29) to 29.4°C at 3:00 (LT 23:00) in Santiago, Dominican Republic (D.R.) during vapor eruption.

(**f**) Magenta square, blue diamond and black dot respectively present the maximum temperature on Dec.29 in Barahona, Las Americas and Punta Cana (All in D.R.) from 1996 to 2012. Considering the maximum temperature on Dec.29 in these three places, 2009 (Y) was the maximum from 1996 to 2012.

(**g**) Port-Au-Prince, Haiti is the station closest to epicenter. However, many temperature data points (S_{1-4}) were missing during the vapor eruption on Dec.29~30, 2009. Satellite data and temperature data were from DU and WU respectively.

The earthquake clouds and geoeruption made four stations of the Dominican Republic (D.R.) temperatures abnormal ("*Definition of Abnormal Temperature*" 1.8; Shou, 2011): Santiago (Red in Fig.40a) got a pulse increase of 5.4°C in one hour on Dec.30 (Fig.40e), and the maximum temperature of Dec.29 in Barahona, Las Americas and Punta Cana was higher in 2009 compared any other years from 1996 to 2012 (Fig.40f). Barahona, Las Americas and Punta Cana had a distance of 145, 290 and 427km to the M7.3 Haiti earthquake respectively. The closer to the Haiti epicenter a downwind station, the higher temperature it recorded (Fig.40f).

Haiti only had the Port-Au-Prince airport to record weather data. Unfortunately, it missed many data points during vapor eruption (Fig.40g), probably because those data were deemed too abnormal. Along the linear cloud, stations Bluefields, Puerto Limon, Kingston, and Guantanamo Bay NAS also recorded a small increase of about 1°C from Dec.29 to 30 (Fig.40a).

2.7.5 The M9.1 Honshu, Japan earthquake

Fig.41 depicts erupting vapors from a swarm of large earthquakes in Japan from Feb.23 to Feb.25, 2011 and their accompanying abnormal temperatures. The largest earthquake was the M9.1 Honshu earthquake on Mar.11. In an area of ±3° around it, 64 big earthquakes happened in March, three of which were ≥ 7. It is difficult and unnecessary to distinguish which vapor originated from which nozzle for all these big earthquakes in images of moderate resolution (4 km/pixel) and low frequency (8 images/day).

However, we can still find the M9.1 epicenter exactly in Geoeruption G_1 at 0:00 on Feb.23 (Yellow edge, Fig.41b). We can also find a big and long straight cloud originating from the M9.1 epicenter southwestward from earlier than 9:00 and persisting to at least 18:00 on Feb.24 and a geoeruption at the same place at 3:00 on Feb.25. The duration of the M9.1 vapor is at least 51 hours from 0:00 on Feb.23 to 3:00 on Feb.25, which is close to the 54-hour duration of the M9 Sumatra earthquake.

During vapor eruption, abnormal temperatures were prevalent. Table 9 describes temperature increases from Feb.22 to the highest record during the eruption on Feb.23~25 from 17 airports around Honshu (Fig.41a). The 17 airports cover an area of 2.06 million square kilometer. The highest increase is 10°C in Tokyo, and the average is 4.9°C.

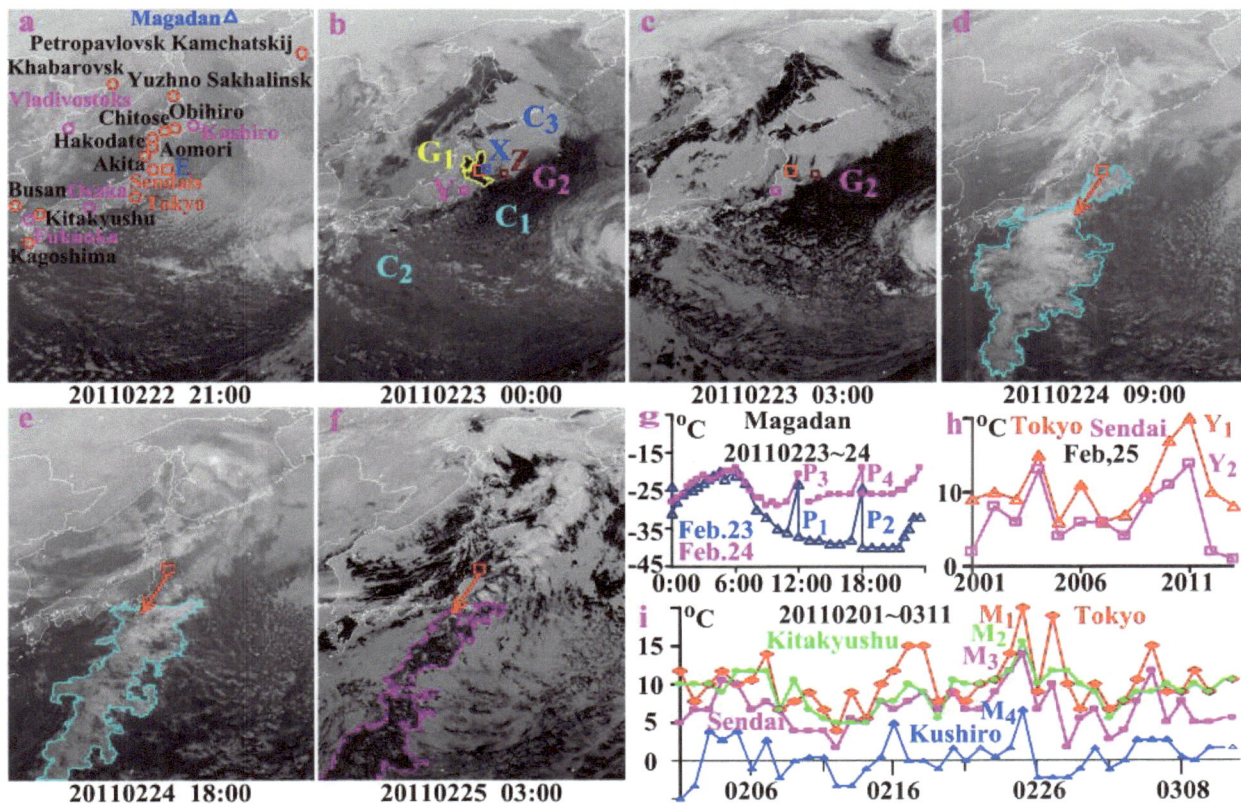

Fig.41 The M9.1 Honshu, Japan earthquake vapor

(**a**) Red square E plots the M9.1 Honshu, Japan earthquake (38.30, 142.37) on Mar.11, 2011. During vapor eruption on Feb.23~25, 2001, one airport displayed a pulse temperature (blue); some airports exhibited the highest daily maximum compared to daily maxima from Feb.1 to Mar.11 (magenta), and some others experienced one or more highest daily maximums compared to daily maximums of comparable dates from 1997 to 2013 (red).

(**b**) Geoeruption G_1 (Yellow edge) appeared in Honshu at 0:00 on Feb.23, 2011 and coincided with the M9.1 epicenter (red square). Blue square X, magenta square Y, and brown square Z respectively plot an earthquake of M7.5 (38.44, 142.84) on Mar.9, M7.9 (36.28, 141.11) on Mar.11, and M7.7 (38.06, 144.59) on Mar.11. C_1, C_2 and C_3 are wave-shaped earthquake cloud, gray earthquake cloud, and heavy gray earthquake cloud respectively. G_2 is another big geoeruption.

(**c**) Geoeruption G_2 became bigger.

(**d-f**) Red arrow emphasizes the southwest direction from the M9.1 epicenter E which coincided with a large, long cloud lasting from 9:00 to at least 18:00 on Feb.24 and a large, straight geoeruption at 3:00 on Feb.25.

(**g**) Temperatures of Magadan, Russia on Feb.23 (blue) and Feb.24 (magenta). P_1, P_2, P_3 and P_4 show four pulse temperature increases: 12.7°C/h (from -36°C at 11:00 on Feb.23 to -23.3°C at 12:00 or LT 22:00~23:00), 13.6°C/h (from -38°C at 17:00 on Feb.23 to -24.4°C at 18:00 or LT 4:00~5:00 on Feb.24), 7.4°C/h (from -28°C at 11:00 on Feb.24 to -20.6°C at 12:00 or LT 22:00~23:00), and 7.1°C/h (from -26°C at 17:00 on Feb.24 to -18.9°C at 18:00 or LT 4:00~5:00 on Feb.25).

(**h**) Y_1 and Y_2 respectively express the daily maximum temperature of Feb.25 among 13 years from 2001 to 2013 in Tokyo and Sendai. The highest daily maximum temperature of Feb.25 coincided with the eruption in 2011.

(**i**) M_1, M_2, M_3 and M_4 respectively reveal the highest daily maximum temperature in Tokyo, Kitakyushu, Sendai, and Kushiro from Feb.1 to Mar.11. They all coincide with the eruption on Feb.25. Satellite images, earthquake data and temperature data were from DU, the USGS and the WU respectively.

The Magadan airport (Blue triangle, Fig.41a) recorded four pulse increases at midnight or in the early morning on Feb.23~24 (P_1~P_4 in Fig.41g). The largest increase was 13.6°C in one hour from -38°C at 17:00 on Feb.23 to -24.4°C at 18:00 (LT 4:00~5:00am on Feb.24).

Other airports (magenta and red circles in Fig.41a) also recorded abnormal temperature increases. For example, in several airports (Fig.41h, e.g. Tokyo and Sendai) if we compare the daily maximum temperature of Feb.25 from 2001 to 2013, Feb.25 of 2001 (during vapor eruption) reached the highest. In several other airports (Fig.41i), the highest daily maximum temperature from Feb.1 to Mar.11, 2011 coincided with vapor eruption on Feb.25.

The above discussion shows that all devastating earthquakes of death toll ≥10,000 from Jan.1, 1990 to Jan.1, 2016 had vapor precursor, i.e. no miss. Simultaneously, each of these discussed vapors predicted a devastating earthquake, i.e. no false warning. The discussion also shows that all discussed earthquakes caused abnormal temperature in the nearby area except for the Rasht earthquake in 1990 which lacked reliable temperature records. Among the nine devastating earthquakes, Shou predicted five in Rashtm, Izmit, Bam, Sumatra (Shou, 2006b), and Kashmir.

2.8 Calculation of Distance and Area

Fig.42 is a schematic diagram for calculating distances and areas. Suppose two points A and B on the Earth have coordinates (A_1, A_2) and (B_1, B_2) respectively (Note: to fit earthquake catalogs, A_1 and B_1: Latitude, A_2 and B_2: longitude). Angle **c**= B_2-A_2.

R= AO = BO = NO = 6365.742km.   Obviously, AC=R·Sin**a**; OC=R·Cos**a**; BD=R·Sin**b**; OD=R·Cos**b**; CD^2= OC^2+OD^2-2OC·OD·Cos**c**= R^2·(Cos^2**a**+Cos^2**b**-2Cos**a**·Cos**b**·Cos**c**). ACBD is a vertical trapezoidal, so AB^2= CD^2+$(BD-AC)^2$= R^2·(Cos^2**a**+Cos^2**b**-2Cos**a**·Cos**b**·Cos**c** +Sin^2**a**+Sin^2**b** - 2Sin**a**·Sin**b**)= $2R^2$·(1-Sin**a**·Sin**b**-Cos**a**·Cos**b**·Cos**c**)   (1)

Therefore,

AB=R·$[2(1-Sin$**a**·Sin**b**-Cos**a**·Cos**b**·Cos**c**$)]^{0.5}$ (2)

In ΔAOB,

AB^2=AO^2+BO^2-2AO·BO·Cos**o** =$2R^2$(1-Cos**o**)   (3)

Comparing (1) and (3),

Cos**o**=Sin**a**·Sin**b**+Cos**a**·Cos**b**·Cos**c**   (4)

ArcAB=R·Arccos(Sin**a**·Sin**b**+Cos**a**·Cos**b**·Cos**c**)   (5)

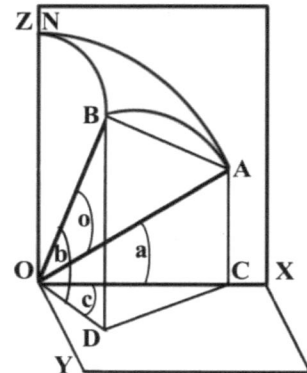

Fig.42 Schematic diagram of distance calculation

Let the center of the Earth be at O and the North Pole N be on OZ. Two points A and B on the surface of Earth have coordinates (A_1, A_2) and (B_1, B_2) respectively (Note: to fit earthquake catalogs, A_1 and B_1: Latitude, A_2 and B_2: longitude). Let A be on Plane ZOX and C be the projection of A on OX. Let D be the projection of B on Plane XOY. **a**=∠AOC, **b**=∠BOD, **c**=∠COD, and **o**=∠AOB.

In computer, this formula can be written as the following in Excel

ArcAB=6365.742*ACOS(SIN(RADIANS(A_1)* SIN(RADIANS(B_1)+COS(RADIANS(A_1)*COS(RADIANS(B_1)*COS(RADIANS(B_2- A_2)) (6)

Also,   AB=6365.742*SQRT(2*(1-SIN(RADIANS(A$_1$)*SIN(RADIANS(B$_1$)-
COS(RADIANS(A$_1$)*COS(RADIANS(B$_1$)*COS(RADIANS(B$_2$-A$_2$))   (7)

Take Fig.29 as an example. A$_1$=-20, A$_2$=60, B$_1$=-30, B$_2$=70. We can obtain 1497.8.km from
(6) for the distance between A (-20, 60) and B (-30, 70). To calculate the length of CD, we use a ruler to compare the lengths of CD and AB in the image, and get 45:125. Thus, the cloud has a real length of 539km.

To calculate the area of a triangle, let a, b, and c be three edges of the triangle, and s=0.5·(a+b+c). Then, this triangle has an area

A=[s·(s-a)·(s-b)·(s-c)]$^{0.5}$ (8)

The above formula is common in high school textbook. Its computer formula is

A=SQRT((s*(s-a)*(s-b)*(s-c))   (9)

A polygon can be divided into many triangles to calculate its area. For example, the thermal abnormal area of Honshu eruption (Fig.41a) can be divided into five triangles (Fig.43). In Column "Triangle", B: Busan, Ka: Kagoshima, Kh: Khabarovsk, Ku: Kushiro, M: Magadan, T: Tokyo, V: Vladivostok. ΔTBKa has three edges TB, BKa and KaT that are arranged by a, b

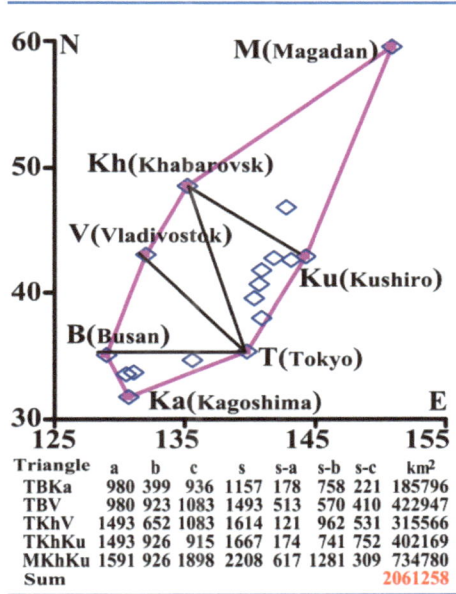

Triangle	a	b	c	s	s-a	s-b	s-c	km^2
TBKa	980	399	936	1157	178	758	221	185796
TBV	980	923	1083	1493	513	570	410	422947
TKhV	1493	652	1083	1614	121	962	531	315566
TKhKu	1493	926	915	1667	174	741	752	402169
MKhKu	1591	926	1898	2208	617	1281	309	734780
Sum								2061258

Fig.43 Thermal abnormal area of Honshu eruption

and c respectively. s=(a+b+c)/2=(980+399+936)/2= 1157 (km). Area A=√[s(s-a)(s-b)(s-c)]= √(1157x178x758x221)=185,796 (km^2). The sum of the five triangles is about 2,060,000km^2.

Chapter 3   Evaluation of earthquake predictions

We have discussed the vapor model and various appearances of earthquake vapor to show how to distinguish earthquake vapor from weather clouds, and how to predict earthquakes from various types of eruptions. We have assessed the "quality" of a prediction by how long a period the earthquake is the only one in a certain area and magnitude, or by how big an area the earthquake is the only one in a certain period and magnitude. In this chapter, we will systematically evaluate the quality of a set of earthquake predictions as far as their accuracy and precision are concerned. That is, the "information content" in a set of predictions. The higher information content, the more valuable predictions are. Specifically, we will introduce a Bayesian method to evaluate a set of Shou's predictions submitted to and signed by the USGS. We will also analyze what caused Shou's prediction mistakes and what compromised the precision of predictions. We will also discuss other precursors.

3.1 Synopsis

An earthquake prediction should consist of a pre-specified window of time, location, and magnitude. A prediction is useful if it is both accurate and precise. The accuracy of a prediction can be verified by comparing the prediction against earthquake records e.g. the catalog of the USGS@2 retrospectively.

The information content (accuracy and precision) of a set of predictions can be evaluated by a variety of statistical methods. In these methods, a test statistic quantified from predictions is compared to its distribution under a null hypothesis (Holliday et al., 2012; Jones and Jones, 2003; Kossobokov et al., 1999; Molchan and Romashkova, 2011; Smyth et al., 2012; Zechar et al., 2010; Zhuang, 2010). Most of these methods require the specification of a stochastic reference model for seismicity. Since most reference models used empirical scaling laws to tie rates of large earthquakes to rates of smaller earthquakes, and are calibrated using data from one region, it is unclear how general these reference models are (Luen and Stark, 2008). Furthermore, evaluation methods that penalize missed predictions can be problematic, especially if predictions were based on incomplete precursor data.

Here, we have chosen to use a Bayesian method proposed by Jones and Jones (2003) to evaluate predictions. For each prediction, whether correct or incorrect, this method calculates a prior probability for the "associated earthquake", the earthquake matching the closest to the prediction. The prior probability (P_{Comb}) incorporates the seismic history of the predicted location (P_C) and potentially increased seismicity after a big shock (P_{RJ}). An information score is then calculated based on the accuracy and the prior probability of the prediction.

Thus, a correct prediction in a seismically active area has a lower positive information score than that in a seismically inactive area; an incorrect prediction in a seismically active area has a more negative information score than that in a seismically inactive area. The information score is then adjusted such that the adjusted information score (*Adj. Score*) would be zero for no skills.

For a large number of independent random predictions, the adjusted scores would be a standard normal distribution according to the central limit theorem. A distribution of adjusted scores significantly different from the standard normal distribution would indicate skills.

Using this statistical method, Shou's set of earthquake predictions submitted to the USGS demonstrates a high level of skillfulness.

3.2 Results

From 1994 to 2001, Zhonghao Shou submitted a total of eighty-six earthquake predictions, including five cancellations, to the USGS office at Pasadena, CA. Each prediction included a location, time, and magnitude window, and was signed and filed by a USGS staff member on the day of submission (Table 10).

A total of sixty-three predictions were independent. Two predictions are considered to be dependent if their location, time, and magnitude windows all had at least partial overlaps, except for the following case: If an earlier prediction was accurate, then the prediction was fulfilled. In this case, even if a later prediction had a time window that partially overlapped with that of the earlier prediction, the two predictions are considered independent.

For two dependent predictions with non-identical time windows, only the earlier prediction was included in the sixty-three independent predictions. In one case, two dependent predictions had identical time windows (Table 10, D1 and D2) and both were excluded from the set of sixty-three independent predictions. The five cancellations were obviously dependent on their original predictions, and were therefore also excluded from the set of independent predictions. In these cases, the original predictions were evaluated.

We first analyzed the set of sixty-three independent predictions (Table 11) to show that they demonstrated a high level of skillfulness. The analysis of the dependent predictions will be presented later to demonstrate that their inclusion or exclusion did not alter our conclusions. For predictions with coarse- and fine-window statements, only the coarsest statements with the largest-size windows were examined.

3.3 Accuracy of predictions

Even though the earthquake catalog of the USGS has errors and the location of epicenter is not just a point, we suppose that the catalog has no errors and the epicenter is a point (instead of an area). For each prediction, we first asked whether it was accurate in magnitude, time, and location windows by searching through the earthquake database. Some earthquakes in the database have a few magnitude values, and in these cases, the largest value is used as the magnitude value of the earthquake as per convention.

The magnitude of an earthquake must be exactly within the predicted window without exception, to be considered a potential hit. The prediction time window was reported in US Pacific time. The timing of an earthquake must occur within the predicted time window without exception to be considered a potential hit. If predicted locations were described in terms of countries and territories, legal boundary (as defined by Google @21) was used to test whether an earthquake is a potential hit.

If predicted locations are described in latitude and longitude values, the precision is rounded to be 0.1 degree. For example, if a predicted latitude window is 35.8~40, then an earthquake at latitude 40.04 (approximated to 40.0) will be considered a potential hit while 40.05 (approximated to 40.1) will not. This convention is also used to calculate the catalog probability Pc (see below). For instance, if the predicted latitude window is 35.8~40, then to calculate Pc, the latitude window for historical earthquakes is set to be 35.75~40.04, not 35.8~40.0. A prediction is considered accurate if there is at least one earthquake that satisfies all three windows ('All' column of "Accuracy" in Table 11). Out of sixty-three predictions, thirty-eight (60%) are accurate.

To statistically evaluate the entire set of predictions, including erroneous predictions, an associated earthquake for each prediction was identified. For an accurate prediction, if multiple earthquakes satisfied the prediction, then the first earthquake is considered the associated earthquake of the prediction, and the prediction is considered fulfilled. If no earthquakes satisfied a prediction, then the prediction is inaccurate and the most closely-matched earthquake with the fewest number of wrong windows was identified to be the associated earthquake (Table 11).

In three cases, predictions missed by a small amount: two had magnitude off by ≤ 0.2 and one had location off by 0.4° ("*" in Table 11). In three other cases, instead of a large predicted earthquake, a swarm of smaller quakes occurred ("#" in Table 11). In 24 out of the total 25 inaccurate predictions, the associated earthquake differed from the prediction in only one of the three windows.

3.4 Calculating the prior probability

To evaluate how skillful Shou's set of predictions is, we calculated the prior probability (Pcomb) for each prediction according to Jones and Jones' (2003) method. This probability has two elements: P_C (catalog probability) and P_{RJ} (aftershock probability). P_C is based on how frequently earthquakes in the predicted magnitude window had occurred in the predicted area in a period equivalent to the length of the predicted time window in the earthquake database (Harrington and Shou, 2005). Specifically, we first calculated P_c (Table 11, column "P_c") for each associated earthquake using the USGS earthquake database spanning from Jan.1, 1990 to Mar. 31, 2012. From the database, we calculated A, the total number of intervals containing exactly the same number of days as the predicted time span. We counted B, the total number of these intervals that contained at least one earthquake whose epicenter is within the predicted area and whose size is within the predicted magnitude range. Then the probability (Harrington and Shou, 2005)

$$P_c = B/A \tag{10}$$

@21 Google map http://maps.google.com/

This quantity effectively represents the chance that a randomly selected time window of the same span as specified in the prediction would harbor an earthquake of the predicted magnitude within the predicted area.

Next, we calculated P_{RJ} for all associated earthquakes (Table 11) to account for the possibility that an associated earthquake may have an increased chance of occurrence as an 'aftershock' of a prior, larger 'mainshock'. The method of calculation was according to Reasenberg and Jones (1994). Based on the USGS earthquake catalog starting from Jan.1, 1990, we looked for potential shocks of larger magnitudes than the associated earthquake in a square area extending $1°$ to the South, North, East, and West of the epicenter of the associated earthquake. The choice of $1°$ ($1°$ latitude corresponds to ~110 km) was based on the spatial distribution of 'aftershocks' (Utsu, 2002): for an M8 and M6 earthquake, aftershocks approximately concentrated in a square of 100 km x 100 km and 10 km x 10 km, respectively, with the epicenter being the center of rectangle. Thus, a $2°x2°$ square would conservatively cover 'aftershocks' of main shocks of at least M8.

Suppose that at least one earthquake larger than the associated earthquake was identified in the $2°x2°$ square centered at the epicenter of the associated earthquake since Jan.1, 1990. Because the area distribution of aftershocks depends on the magnitude of the mainshock, we examined whether or not the associated earthquake indeed fell into the 'aftershock' regions of these larger earthquakes as defined by Utsu's formula (Utsu, 2002). If so, then all those larger earthquakes were considered potential mainshocks for the associated earthquake.

For the eleven cases, preceding with potential main shocks (Table 11, "1" in Column "Af"; Table 12), we calculated P_{RJ} for each potential 'mainshock'. P_{RJ} is the probability of a 'mainshock' having at least one 'aftershock' with the magnitude and time windows specified by the prediction. The formula provided by Reasenberg and Jones (1994) states that the rate of aftershock of magnitude M or larger is:

$$r(t, M) = 10^{a+b(Mm-M)}(t+c)^{-d} \qquad (11)$$

where Mm is the magnitude of the mainshock, t is the time after mainshock, and a, b, c, and d are constant parameters set to be -1.67, 0.91, 0.05, and 1.08, respectively (Reasenberg and Jones, 1994). Thus, the expected number of 'aftershocks' of magnitude between M_1 and M_2 in a time window of $[t_1\ t_2]$ is

$$\lambda(M_1, M_2, t_1, t_2) = \int_{M_1}^{M_2} \int_{t_1}^{t_2} 10^{a+b(Mm-M)}(t+c)^{-d}\, dt\, dM$$

$$= \frac{10^{a+bMm}}{1-d}\left((t_2+c)^{1-d} - (t_1+c)^{1-d}\right) \frac{e^{(-b\ln 10)M_1} - e^{(-b\ln 10)M_2}}{b\ln(10)} \qquad (12)$$

When the upper-bound of magnitude was not specified in a prediction, M_2 was set to be $+\infty$. The expected number of 'aftershocks' λ was used to calculate P_{RJ}, the probability of having at least one 'aftershock' satisfying the predicted time and magnitude windows, assuming a Poisson distribution with parameter λ:

$$P_{RJ} = 1 - e^{-\lambda(M_1, M_2, t_1, t_2)} \qquad (13)$$

If an associated earthquake could be an 'aftershock' for multiple main shocks, the main shock with the largest P_{RJ} was used to make our statistical tests most conservative (Tables 11, 12). For predictions not associated with any main shocks, $P_{RJ} = 0$.

Finally, a combined prior probability was calculated as $P_{Comb} = P_C + P_{RJ} - P_C P_{RJ}$ (Jones and Jones, 2003 Table 11, Column "P_{Comb}"). This calculation was based on the assumption that the probability $P(A\ or\ B)$ can be expressed as $P(A)+P(B)-P(A\ and\ B)$ and the assumption that P_C and P_{RJ} are independent.

3.5 The set of sixty-three independent predictions demonstrate statistically significant skills

To test whether the set of predictions was better than random guesses, we used the method of Jones and Jones (2003). A prediction that an earthquake is going to occur within a specified time, magnitude, and location windows can be accurate or inaccurate. If the prior probability of a prediction is p, then an accurate prediction will get a score of $-\ln p$ (a positive number) while an inaccurate prediction will get a score of $\ln(1-p)$ (a negative number) (Jones and Jones, 2003). If the predictor has no skills, then the expected score would be $-p\ln p+(1-p)\ln(1-p)$.

Subtracting a score with the expected score gives rise to an adjusted score (Table 11, Column "Adj Score") whose mean will be zero if there is no skill (Jones and Jones, 2003). The variance of each score (Var.) is

$$Var = p(1-p)[\ln(p(1-p))]^2 \qquad (14)$$

(Table 11, Column "Var"). For a large enough set of independent predictions (greater than fifteen), the result should be a standard normal distribution if there is no skill (Jones and Jones, 2003). Since all predictions are independent, the scores and variance were summed over all predictions.

We obtained a total score of 15.3 and variance of 28.8. Thus the Z-score for the set of sixty-three independent predictions was 2.84($=15.3/\sqrt{28.8}$). The probability that by chance, a person with no skill could obtain a score at least as high as the observed total score is the P-value. P-value of ≤ 0.05 is used as a criterion for statistical significance. According to the Z-score, The P-value for this set of sixty-three predictions is 0.002, calculated as a two-tailed test from the standard normal distribution. Thus, the set of predictions outperforms random guesses in a statistically highly significant fashion. Moreover, the accuracy of this set is 60%.

3.6 Dependent predictions did not affect performance

Out of the twenty-three dependent predictions, five were cancellations (D6, D10, D14, D16 and D17 in Column "No." of Table 10) of five respective earlier predictions (#39, #49, D13, D15 and #55 in Column "D" of Table 10) when new or higher-resolution precursor data became available. All these five cancellations, three of which announced before half of the original prediction time-window had elapsed, were correct. Even though weather forecasts also often revise predictions, we disregarded the cancellations and included the three erroneous original predictions in the set of sixty-three independent predictions and the two erroneous prediction revisions in the set of dependent predictions to set the maximum level of rigor.

Of the eighteen remaining dependent predictions, nine were accurate. We calculated the combined prior probability P_{comb} for all eighteen dependent predictions. The total score was 0.4 with a standard deviation of 8.6. This, in contrast with the set of independent predictions, did not show statistically significant skill (P-value= 0.45). However, when combined with the sixty-three independent predictions, the total score was 15.7 with a standard deviation of 6.1. The set of eight-one predictions, eighteen of which are dependent, still showed a high level of skill, with a Z-score of 2.6 and a P-value of 0.005 and an accuracy of 60%.

3.7 Problems of evaluating predictions

We completely followed Jones and Jones' evaluation method to be conservative. Still, the method has two problems. First, the earthquake database of the USGS contains 'aftershocks', so it is unnecessary to recalculate 'aftershock' probabilities. Moreover, the term 'aftershock' does not have a scientific definition (Utsu, 2002). Consider an earthquake B followed earthquake A. If A>B, then B is traditionally considered an aftershock of A. If A<B, then A is traditionally considered a foreshock of B. However, when A=B (e.g. two M6.6 Iceland earthquakes: one on Jun.17, 2000 and the other on Jun.21), then the relationship between the two is no longer clear.

Moreover, Reasenberg and Jones' (1994) 'aftershock' formula is not universal. For example, it gives the M6.8 Olympia, Washington State earthquake on Feb.28, 2001 an aftershock probability of 100% for magnitude ≥4 in three months from Feb.28 to Jun.1 in an area of about 0.2° degree east, west, south, and north of the epicenter. However, no such 'aftershock' occurred even in a larger area of 3° degree east, west, south, and north of epicenter. Therefore, Jones and Jones' (2003) method can significantly enlarge aftershock probabilities. Furthermore, most other evaluation papers do not adopt aftershock probabilities. When we removed all aftershock probabilities, we got a P-value of **0.001** and **0.002** for the 63 independent predictions and the 81 predictions respectively.

We supposed that earthquake data of the USGS were without errors, and that an earthquake epicenter hit a point, not an area. In reality, all earthquake data have errors (Table 6) and even misses (Table 5, 7). An earthquake hits an area, not a point. For example, the M6.1 Afghanistan earthquake on Feb.4, 1998 destroyed an area of 36.83~ 37.31N, 69.5~70.11E@17. This stringent assumption caused two small errors in magnitude and a small error in area. They together caused a minus score of -2.645 (* in Table 11).

3.8 Causes of prediction errors

First, poor quality of satellite images is a major cause of erroneous predictions. Although the Bam cloud clearly pinpointed the epicenter (Fig.6,7), the Canada cloud did not (Fig.25) and thus caused a miss-prediction in location initially (No.61 prediction in Tables 10 and 11) although the modified prediction with enlarged area was excellent (D23 in Table 10). This mistake in location prediction is caused by temperature grey scales of satellite images being artificially limited to a small range, thus severely limiting the visibility of the hottest spot corresponding to the epicenter, as discussed in 'Problems caused by satellite images' (Chapter **2.5**). This satellite image problem and Shou's previous lack of experience in estimating epicenters caused 8 area mistakes and a minus score of -1.925 (0 in L of Table 11).

Second, low resolution satellite images, discussed also in 'Problems caused by satellite images' and exemplified by the Indian cloud (Fig.29b), caused 4 mistakes in magnitude windows and a minus score of -0.838 (# in Table 11).

Third, Shou's previous lack of experience in estimating the lag between eruption and earthquake (See Fig.33a) caused 11 mistakes in time windows and a minus score of -5.616 (0 in T of Table 11).

Finally, in the section above (3.7), we discussed a minus score of -2.645 (* in Table 11) due to our stringent assumptions that the USGS data have no errors and that an epicenter is focused to a point instead of occurring over an area (both of which are highly unlikely).

The four problems together not only caused all negative scores, but also reduced the scores of correct predictions by enlarging their windows. For example, No.24 Afghanistan earthquake prediction had a large area window of 25~41N, 53~105E because the cloud was near the edge of the image and the image did not have coordinate grids (Fig.30b). We have discussed that the image from IODC satellite of Dundee (2.2.6. Linear earthquake cloud) alone could reduce the area window of Shou's prediction by 20 times, which would reduce the Pcomb to about 0.215 and increase the score from 0.77 to 1.40.

The above analysis points at the main causes of prediction errors and shows that the P-value 0.001 of Shou's 63 independent predictions is a conservative estimate. This P-value does not represent the usefulness of the vapor precursor, but rather Shou's skills of using the precursor under the limitations discussed above.

3.9 Values of the Vapor Precursor

Even though Shou's Bam prediction has superb quality, with a probability of being correct by chance close to zero, the prediction could still have been made far better based on the vapor precursor. The nozzle of the Bam cloud coincided with the epicenter exactly, but Shou's lack of experienced made him extend the area to AB (Fig.7a), far bigger than the nozzle.

Moreover, Shou's initial magnitude window was ≥5.5 which was posted in a hurry because of a four-day delay in his reviewing the images. After investigating the M5.6 Southern Iran cloud (Fig.34), he increased the magnitude to ≥6.5, closer to M6.8 of the USGS report, to warn Iran people, but the earthquake had just happened.

Furthermore, Shou predicted a 60-day time window. However, the second Bam cloud (Fig.7e) and the second temperature peak (Fig.7c) on Dec.25, 2003 coincided with the earthquake on Dec.26. Thus, the value of vapor precursor is far better than Shou's Bam prediction.

When we understand vapor precursor sufficiently, it can even be used as a complementary method to check the quality of earthquake data, e.g. the precision of epicenters (Fig.7a,e, 8c, 11b,g, 20a,b,d, 22, 33, Table 2, 5, etc) and data misses (Table 5).

3.10 Other prediction methods

Geller et al. (1997) criticized, "Thousands of observations of allegedly anomalous phenomena (seismological, geodetic, hydrological, geochemical, electromagnetic, animal behavior, and so forth) have been claimed as earthquake precursors, but in general, the phenomena were claimed as precursors only after the earthquakes occurred. The pattern of alleged precursors tends to vary greatly from one earthquake to the next, and the alleged anomalies are frequently observed at only one point, rather than throughout the epicentral region. There are no objective definitions of "anomalies," no quantitative physical mechanism links the alleged precursors to earthquakes, statistical evidence for a correlation is lacking, and natural or artificial causes unrelated to earthquakes have not been compellingly excluded." These are reasonable points although it was illogical to claim that "Earthquakes Cannot Be Predicted". We will discuss problems associated with other prediction methods and their precursors. I will give examples of predictions based on those precursors and discuss problems with the precursors.

3.10.1 Haicheng Earthquake Prediction based on geodetic and foreshocks

The Haicheng Earthquake Study Delegation (1977) claimed the Haicheng prediction as "an extraordinary achievement" and "the first major shock to have been accurately predicted anywhere in the world". Started in 1970, the Chinese scientists adopted geodetic to find an uplift of about

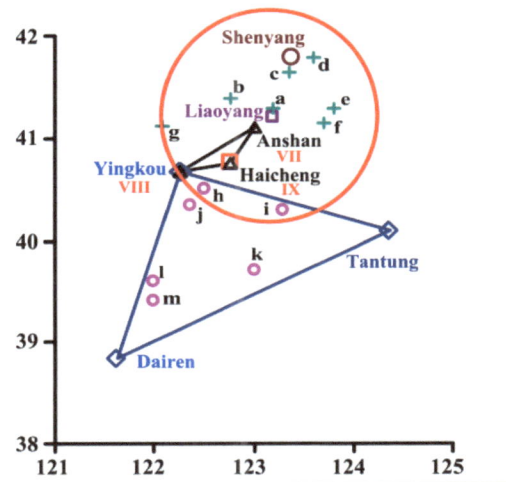

Fig.44 Problems of the Haicheng prediction
Yingkou-Dairen-Tantung (Blue triangle) shows a predicted area of an M5.5-6 earthquake in the first 6 months of 1975 by the State Seismological Bureau (SSB) in a conference on Jan.13, 1975. In a claimed imminent-term prediction by city committees in emergency meetings at LT 14:00 on Feb.4 or after, Haicheng-Yingkou-Anshan (Black triangle) were the predicted area. Red square plots an M7.3 earthquake in Haicheng at LT 19:36 on Feb.4. Red numbers IX, VIII and VII reveal damage intensity. Purple square plots the largest foreshock of magnitude 4.8 at Liaoyang before LT 8:00 on Feb.4. This shock was supposed to be the main evidence for the imminent-term prediction. Red circle represents an area of 100km radius centering Liaoyang. This circle includes Yingkou (99km from Liaoyang). Letters a through m plot cities Dengta, Liaozhong, Sujiatun, Dongling, Benxi, Nanfen, Panshan, Dashiqiao, Xiuyan, Gaizhou, Zhuanghe, Wafangdian, and Pulandian respectively. Information of the predictions and the earthquakes are from the (Haicheng Earthquake Study Delegation, 1977). Coordinates of plotted cities are from Atlas of the World, WU and the Google maps

30 mm in Liaoning peninsula and made a short-term prediction of an M5.5~6 earthquake in

Yingkou-Dairen-Tantung in the first 6 months of 1975. The scientists further made an imminent-term prediction relying on an M4.8 foreshock at Liaoyang before LT 8:00 on Feb.4. Emergency meetings for evacuation began in both government and public organization of Haicheng, Yingkou and Anshan at or after LT 14:00. Then, emergency shelters, medical teams, etc were prepared. At LT 19:36, an M7.3 gave Haicheng, Yingkou and Anshan a damage grade IX. VIII and VII respectively (Fig.44).

However, Geller et al. (1997) denied this achievement. They rejected seismological precursor. They pointed out that the reported few casualties due to prediction success contradicted with a later report of 1,328 deaths and 16,980 injured people (Quan, 1988). Geller suspected that the political atmosphere of China (culture revolution) might have generated false information driven by political ideology.

"Foreshock", "aftershock", and "mainshock" do not have scientific definitions (Utsu, 2002). Moreover, many large earthquakes, such as the Tangshan earthquake, the Northridge earthquake, and the Bam earthquake, do not have "foreshocks", so "foreshock" by itself does not predict some major earthquakes. On the contrary, for a period of time in 1997~1998, daily shocks increased dramatically to hundreds per day in Mammoth, a small town in Northern California (Fig.45). Because of this phenomenon, many people urged Shou to predict a big earthquake. He declined because there was no strong earthquake cloud. The fact that the highest magnitude was 5.1 (moderate) among all shocks in Fig.45 shows that high density foreshocks are not always followed by a big earthquake. These observations contradict "There is little question that foreshock activity was the most important observation defining the impending earthquake prediction in both space and time" for imminent prediction (Haicheng Earthquake Study Delegation, 1977).

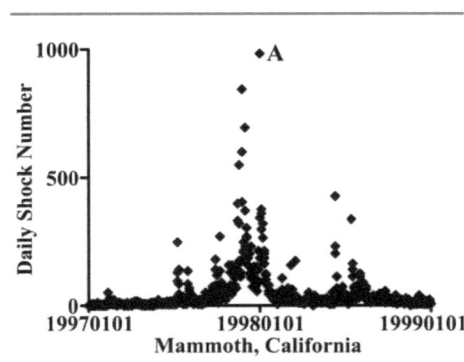

Fig.45 Swarm of shocks in Mammoth
Mammoth (37.3~37.8, -119.1~-118.6) is a small town in Northern California. In 1997 and 1998, news widely reported daily shock increases there and worried about a big earthquake destroying a source of drinking water. Point A shows 982 shocks on Jan.2, 1998. Earthquake data are from the USGS.

There were two predictions. In the short-term prediction on Jan.13, 1975, the predicted area was Yingkou-Dairen-Tantung (blue triangle in Fig.44), which did not include Haicheng. The M4.8 foreshock at Liaoyang was claimed to narrow the imminent-term area to Haicheng-Yingkou-Anshan (black triangle in Fig.44), but Liaoyang itself and its major vicinity (red circle in Fig.44) were excluded without a good reason. Moreover, no resident witnessed the prediction although at least 300,000 people lived in the proclaimed evacuated area. In addition, the foreshock happened at LT 8:00 on Feb.4, the emergency meetings started at LT 14:00, and the devastating earthquake occurred at LT 19:36. Is it possible to evacuate 300,000 people in such short time and a cold winter? The imminent-term prediction was too magic to be true, but the short-term prediction, even though it was wrong, still had value since no prior predictions came close.

3.10.2 Tangshan Earthquake Prediction based on geodetic

Relying on geodetic measurement, the Chinese scientists found an average uplift of 19 mm per year southeast near Tangshan between 1967~1968, and an uplift of 24.1 mm northwest near Tangshan between 1968~1969. They gave an official warning in 1970 that there would be a big earthquake in the Beijing-Tianjin-Tangshan area, even though Tangshan itself did not deform much. After 1970, the deformation in Tangshan even reduced (Xie and Huang 1987, Zhang 1981). If the Chinese scientists had started the measurement after 1970, they might have missed the data. The Tangshan earthquake also did not have a foreshock (Zhang, 1981), so foreshock would miss some earthquakes too. Although the prediction did not have a defined time window and the M7.8 Tangshan earthquake killed 242,769 people in 1976, the prediction is meaningful because it proved that earthquakes can be predicted.

3.10.3 Los Angeles Earthquake Prediction based on geodetic

Began in 1996, a team of ten scientists from the both NASA (National Aeronautics and Space Administration) and USGS set up an array of GPS (Global Positioning System) sensors, linked to satellites, to monitor Los Angeles. On Aug.3, 1999, they predicted to the public that the "next major quake" would be in "Los Angeles".

A woman from Northridge was too nerves to sleep in the midnight and requested Shou to comment on the prediction. He immediately replied that the prediction would fail. First, geodetic was not a short term precursor. Chinese scientists predicted the Tangshan earthquake in 1970, but it happened in 1976. Second, satellite images showed two hot places: one around Palm Spring, Landers, and surrounding area, and the other near mid border between California and Nevada. The weather reporter of ABC TV reported 109°F (42.8°C) in Palm Spring as the highest in Southern California, and supported his theory.

To reply to many anxious people, Shou posted an essay "California Earthquake Situation Analysis" on Aug.10 with an image (Appendix12, Fig.32b without plotting Los Angles). In the essay, he predicted that the next major quake would not hit Los Angeles; instead it would strike one of the two hot spots in the image. The M7.4 Hector Mine earthquake on Oct.16, 1999 showcased his success (**2.3.2** "Earthquake cloud and geoeruption mixture"). Moreover, no major quake has hit Los Angeles so far.

3.10.4 Geodetic Precursor

The Haicheng prediction and the Tangshan prediction show that the geodetic precursor works on a large area and is more suitable for long-term predictions (about 2~8 years). The fact that the deformation around Tangshan reduced after 1970 and Tangshan itself did not deform much (Xie and Huang 1987, Zhang 1981) shows that geodetic may miss prediction. Flood can affect tilt data (Haicheng Earthquake Study Delegation, 1977). Human activity can affect geodetic data

(Clarke, 2001). Therefore, geodetic may have false warning. Clarke (2001) reported that the Los Angeles prediction cost multimillion dollars, even though the prediction was wrong.

On the other hand, the vertical wall of the earthquake monitoring well in No.10 Tangshan high school became sloped 4 days before the quake (Cai et al., 1987). The tilt trajectory at Shihpengyu Observatory, Yingkou, stopped due to drastic movements 16 hours before the Haicheng earthquake (Haicheng Earthquake Study Delegation 1977). Thus, geodetic may fortuitously work for making short-term predictions, but in general it is difficult to determine the time window and the epicenter.

3.10.5 Parkfield Earthquake Predictions

The USGS designated Parkfield (about 36, -120.5) to be a "characteristic" place in San Andreas Fault where big earthquakes had happened with a cycle of 22 years (Thatcher 1992). This hypothesis relied on six big earthquakes in 1857, 1881, 1901, 1922, 1934 and 1966 there. In Apr.1985, the USGS predicted an earthquake of magnitude about 6 near Parkfield before 1993. In Sep.1985, the USGS and the California Division of Mines and Geology jointly set up the most densely and comprehensively instrumented earthquake zone in the world. In both 1934 and 1966, a magnitude 5.0 foreshock preceded the magnitude 6 mainshocks by about 17 min (Langbein, 1992). Thus, foreshock was supposed to predict next big earthquake.

At UTC 5:28 on Oct.20, 1992, an M4.7 earthquake occurred near Parkfield. At 5:46, the USGS predicted an earthquake of magnitude ≥ 6 in Parkfield within 72 hours, but nothing happened (Langbein, 1992).

At 12:25 on Nov.15, 1993, an M4.9 earthquake hit Parkfield. Soon, the USGS predicted another earthquake of magnitude ≥ 6 in Parkfield within 72 hours, but nothing happened again. This experiment rejects the "characteristic" hypothesis and shows that foreshock can generate false warning.

When the USGS issued the second prediction, Shou was in Southern California. Many people asked him about the prediction. He answered that with a probability of 95% the prediction would be wrong. With variations in inter-earthquake interval as high as 10 years, the cycle did not exist in a manner precise enough for short-term predictions. The "characteristic" model was only a variant of foreshock.

Fig.46a shows all big earthquakes in Parkfield and the two erroneous predictions by the USGS. G indicates an M6 earthquake on Sep.28, 2004. Earthquake G also has no foreshock. Therefore, foreshock is neither necessary nor sufficient for predicting main shocks.

Fig.46b reveals a temperature increase of 6°C in Fresno, Hanford, Visalia, Bakersfield and Porterville from 30.6~31.7°C on Jun.12, 2004 to 36.7~37.8°C on Jun.16, which was the highest temperature in the five cities in Jun. 2004. This occurred 104 days prior to the M6 earthquake. By contrast, temperatures in Los Angeles were normal.

Fig.46c-h depict an earthquake cloud from Southern California eastward on Jun.16-17, 2004. Its mass suggests a magnitude about 6. Opposite of its moving direction points toward the nozzle. The time and location of the cloud are consistent with the abnormal temperature in

Fig.46b. Earthquake data reveal the M6 Parkfield earthquake on Sep.28 being the only one of magnitude ≥6 in Southern California for more than eleven years since the cloud on Jun.16, 2004. The high coincidence among the earthquake, the cloud, and the abnormal temperature suggests the M6 Parkfield earthquake being associated with the earthquake cloud.

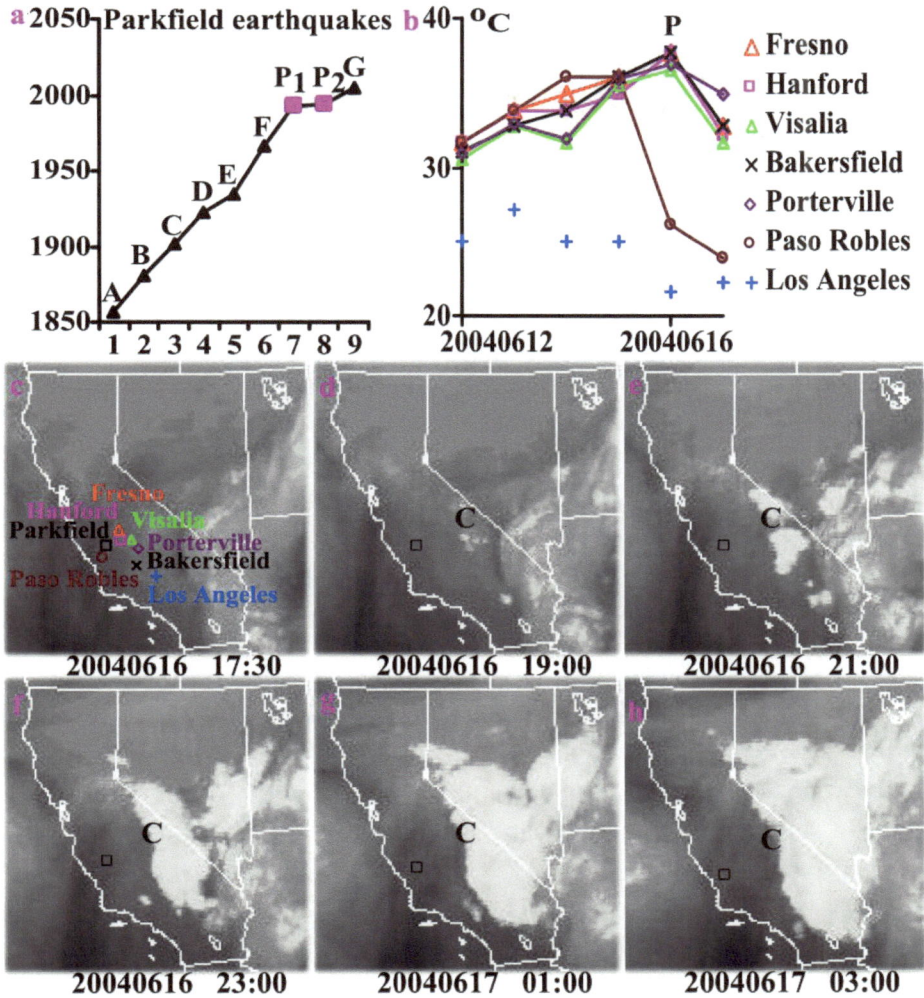

Fig.46 Parkfield earthquake prediction

(a)A, B, C, D, E, F and G respectively show big earthquakes in Parkfield (36, -120.5) and its vicinity in Southern California on Jan.9, 1857; Feb.2, 1881; Mar.3, 1901; Mar.10, 1922; Jun.6, 1934; Jun.28, 1966 and Sep.28, 2004. P_1 and P_2 are predictions by the USGS for an earthquake of magnitude ≥6 in Parkfield within 72 hours on Oct.20, 1992 and Nov.15, 1993 respectively.

(b) Red triangles, magenta squares, green triangles, black crosses, purple diamonds, brown circles, and blue pluses respectively represent daily maximum temperatures of stations in Fresno, Hanford, Visalia, Bakersfield, Porterville, Paso Robles, and Los Angeles on Jun.12~17, 2004. P indicates daily maximum temperature of 36.7~37.8°C in Fresno, Hanford, Visalia, Bakersfield and Porterville on Jun.16, 2004. This temperature was the highest in June 2004. Daily maximum increased 6°C in the five cities from Jun.12 to Jun.16.

(c) Black square plots Parkfield. Locations of cities in **b** are plotted.

(**d-h**) An earthquake cloud C appeared above Southern California at 19:00 and became bigger and bigger until at least 3:00 of next day. Earthquake data, temperature data and satellite data respectively were from the USGS, the WU and the NOAA,

3.10.6 Wenchuan Aftershock Prediction

After the M8 Wenchuan earthquake on May 12, 2008, the SSB issued a prediction for an aftershock of magnitude ≥6 on May 19~20. However, no earthquake of magnitude ≥5.3 occurred in an area of ±5° around the M8 epicenter until an M6 quake on May 25 that Shou predicted (Fig.39f). This comparison reveals that earthquake vapor, not aftershock, is valuable for earthquake prediction.

3.10.7 L'Aquila "No danger" Earthquake Prediction

L'Aquila is a medieval city in middle of Italy. In the first three months of 2009, a swarm of small shocks made people there panic. Government official Bernardo De Bernardinis, then vice-director of the Department of Civil Protection, organized a press conference on Mar.31, 2009. The press conference was tasked with assessing the risks of earthquakes and represented a committee consisting of six scientists: Enzo Boschi, then-president of Italy's National Institute of Geophysics and Volcanology in Rome; Franco Barberi, at the University of 'Rome Tre'; Mauro Dolce, head of the seismic-risk office at the National Department of Civil Protection in Rome; Claudio Eva, from the University of Genova; Giulio Selvaggi, director of the INGV's National Earthquake Centre in Rome; and Gian Michele Calvi, president of the European Centre for Training and Research in Earthquake Engineering in Pavia (Hall, 2011). The press conference assured the public that they were in "No danger".

People were accustomed to leaving homes when feeling a little shock. After the press conference, many people decided to stay. For example, Vittorini, a 48-year-old surgeon, persuaded his wife Claudia and his nine-year-old daughter Fabrizia to stay home with him when a shock (M4.2) appeared on the evening (UTC 20:48) of Apr.5, 2009. By contrast, old people left homes. On next early morning (UTC 1:32), an M6.3 earthquake killed Vittorini's wife and daughter and many others (Hall, 2011).

In June 2010, Public prosecutor Fabio Picuti charged the seven for manslaughtering because local residents had made decisions based on the "No danger" announcement. The American Geophysical Union and the American Association for the Advancement of Science (AAAS) issued statements in support of the Italian defendants. In an open letter to the president of Italy, Giorgio Napolitano, signed by more than 5,000 members, the AAAS criticized local prosecutors for charging the men for failing "to alert the population of L'Aquila of an impending earthquake" even though pinpointing the time, location and strength of a future earthquake "in the short term remains, by scientific consensus, technically impossible" (Hall, 2011). In Oct. 2012, a judge ignored the open letter and sentenced the seven to six years in prison (Nosengo, 2012).

In Nov. 2014, an appeals court overturned the six-year prison sentence for the six scientists and reduced the sentence for the government official to two years because "the scientists' attorneys argued that no clear causal link had been proven between the meeting and the behaviour of the people of L'Aquila. They also argued that the scientists could not be held accountable for De Bernardinis's reassuring statements, and that the seismologists' scientific opinions were ultimately correct." "The only useful thing that can protect us from earthquakes is the seismic hazard map of a country," However, many L'Aquila citizens waiting outside the courtroom reacted with rage, shouting "shame" (Abbott and Nosengo, 2014).

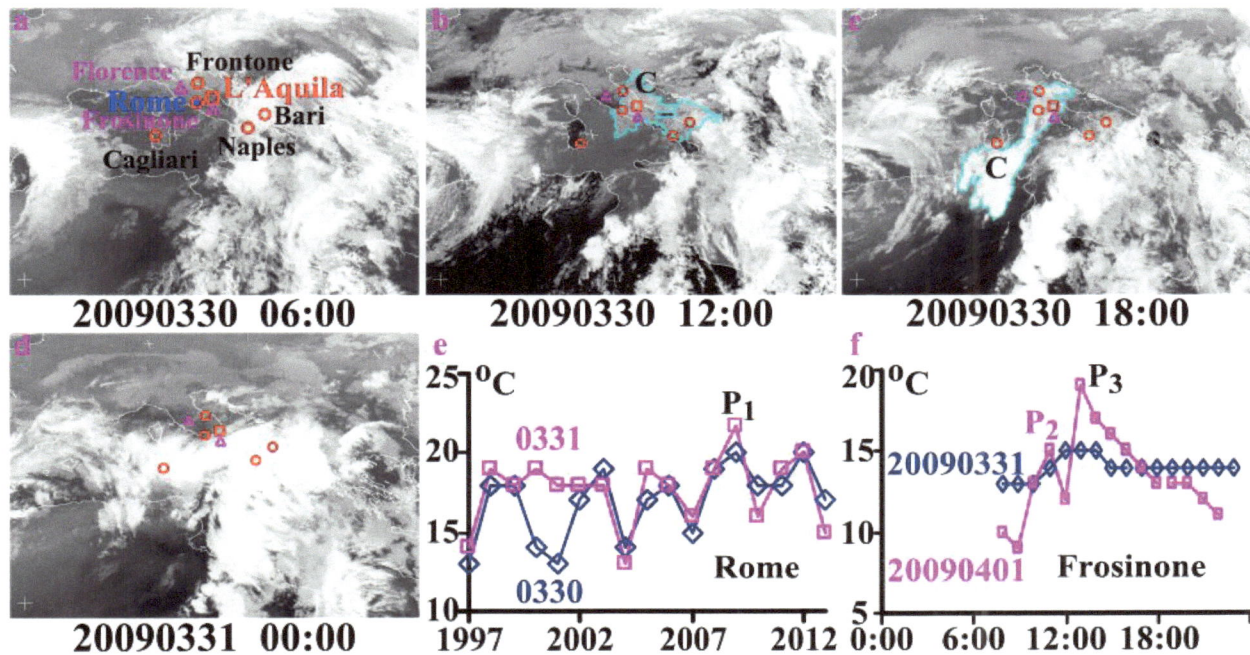

Fig.47 L'Aquila "No danger" earthquake prediction
(**a**) Red square plots the M6.3 L'Aquila epicenter (42.33, 13.33) on Apr.6, 2009. Red circles, and magenta triangles respectively plot cities having one or more highest daily maximum temperatures on Mar.29~Apr.1, 2009 among the records on Mar.29~Apr.1 of 1997~2013, and cities having pulse temperatures. Those temperatures were abnormal.
(**b-c**) Cyan edge reveals an earthquake cloud from L'Aquila traveling southwestward from 12:00-18:00 on Mar.30. The cloud just covered above cities. They were in an area of about 163700km^2. The highest temperatures there increased 3.7°C on average from Mar.28.
(**d**) The earthquake cloud mixed with weather cloud northeastward at 0:00 on Mar.31.
(**e**) Blue diamonds and magenta squares respectively depict daily maximum temperatures in Rome on Mar.30 ("0330") and 31 ("0331") from 1997 to 2013. P_1 shows the temperatures in 2009 to be the highest in records.
(**f**) Blue diamonds and magenta squares respectively reveal hourly temperatures in Frosinone on Mar.31 and Apr.1. P_2 and P_3 show two pulse temperatures. Images, earthquake data and temperature data were from DU, the USGS and WU respectively.

We sympathize with the 309 victims and the committee, but are sorry for their ignoring the Earthquake Vapor Theory (Harrington and Shou, 2005). The United Nations (UN) published it and shared it with its all member states, including Italy, USA, and UK during the 42nd Session of Scientific and Technical Subcommittee in Vienna on Feb.21~ Mar.4, 2005. Furthermore, Shou's theory has already been patented (Patent US 8,068,985B1, Shou 2011). Therefore, it is better to answer two simple questions than to claim "impossible" by consensus. First, how can the Bam cloud appear from and insist at Bam for 26 hours? Second, how can Shou's Bam prediction be so successful?

For the L'Aquila earthquake, there was a big earthquake cloud (Cyan edge in Fig.47b~c) and abnormal temperatures (Fig.47e-f) on Mar.30~31, which were enough to warn people, or at least not denying the possibility of an earthquake. The L'Aquila earthquake cloud appeared at 12:00 on Mar.30, 2009, and rushing southwestward against normal wind at 18:00. It reached Rome on Mar.30 and 31 when daily maximum temperatures reached the highest among the same dates in 17 years from 1997 to 2013 (Fig.47e). Simultaneously, pulse temperatures appeared in Frosinone etc (Fig.47f). In an area of about 163700km^2 (Fig.47a), where the cloud reached, daily maximum temperatures increased 3.7°C on average. In addition, a seismic hazard map is not useful. For example, it made two false warmings in Parkfield (Fig.46a), and missed big earthquakes in Haicheng, Tangshan, Northridge, Bam, L'Aquila and so on.

3.10.8 Foreshocks and Aftershocks

Gerstenberger et al. (2005) claimed "Real-time forecasts of tomorrow's earthquakes in California" using foreshocks and aftershocks, but they have not made a real prediction yet. We have discussed that "foreshock", "aftershock", and "mainshock" do not have scientific definitions (Utsu, 2002), and thus it is impossible to distinguish between foreshocks and non-foreshocks. Further, as we have seen in Mammoth (Fig.45), "foreshocks" are not always followed by big shocks. In contrast, the Tangshan earthquake, the Bam earthquake, and the Northridge earthquake had no foreshock. Among Parkfield earthquakes (Fig.46a), some big earthquakes had foreshocks while others did not. The M6.8 Olympia earthquake (47.14,-122.72) on Feb.28, 2001 had neither moderate foreshock, nor moderate aftershock in an area of ±3° around the epicenter and a period of ±100 days. In short, foreshocks are not reliable earthquake precursors. "Foreshocks" are neither necessary nor sufficient as an earthquake precursor. Even using main shocks to predict aftershocks is difficult.

Fig.28 shows four linear clouds that predicted four earthquakes of magnitude 3.2, 4.2, 4.9 and 4.8. According to foreshock and aftershock hypothesis, the M3.2 and the M4.2 are foreshocks, the M4.9 is the mainshock, and the M4.8 is an aftershock. However, the four earthquakes were the only ones of magnitude ≥3 in the direction where the clouds came from and in the area (38~39, -119~-118) within 310 days from Apr.21, 1997 to Feb.24, 1998. The high coincidence between the four clouds and the four earthquakes suggests the four earthquakes not being foreshocks, mainshock, and aftershock, but all independent.

3.10.9 Radon

Many scientists suppose Radon (Rn), a colorless, odorless, tasteless noble gas and a decay product of Uranium or Thorium, to be an earthquake precursor. Wang et al.(1980) reported Radon anomalies in ground water within a distance of about 360km for 5 years before the M7.2 Sungpan-Pinwu earthquake in 1976. Thus, Radon does not work for short term predictions. Wang et al. (1980) also reported pulse increases of Radon appearing in six springs around the epicenter. However, three others including the one in the epicenter area did not. Chinese scientists submitted four charts of Radon anomalies for the Haicheng earthquake to the Haicheng Earthquake Study Delegation (1977), but the Delegation rejected them all.

Like the Chinese scientists, Silver and Wakita (1996) proposed 'anomalies' for some precursors including Rn before the Kobe earthquake in 1995, and the Izu-Oshima earthquake in 1978. On Jan.14, 2014, Shou asked Wakita three questions: (1) Do you have a model to explain how an earthquake produces Radon? (2) You mentioned "Precursory anomalies began 3 months before the event" for the Kobe earthquake. However, I cannot find them from the curves of 'Strain', 'Rn' and 'Cl'. May you tell me your definition for "Anomalies"? (3) You wrote, "The main shock produced coseismic discontinuous change in all indictors" for the Kobe earthquake. I do find discontinuous change in two Strain curves, but not from the Strain curve of the Izu-Oshima earthquake. So, the discontinuous change works only for the Kobe earthquake, does it not? Wakita replied on Jan.16 that he had retired, and "I am sorry that I have no reply to your questions".

Shou's three questions are fundamental. It is important to know how an earthquake produces a precursor. However, no paper addressed how earthquakes could produce Rn anomaly. Geller et al. (1997) criticized claims of 'anomaly' without a scientific definition, which was another common problem. Thus, Rn cannot serve as an earthquake precursor.

3.10.10 VAN predictions based on SES

VAN is an abbreviation of a group of Greek authors: Varotsos, Alexopoulos, Nomicos etc. They proposed that an impending earthquake will release "seismic electric signals" (SES) with a fixed direction and within 11 days before the quake. Thus, SES was supposed to predict earthquake location and time. Seventeen stations were set up in Assiros, Glyfada etc. Each station had an array of dipoles underground with various lengths e.g. 47.5m, 70m, etc and two directions: EW and NS to receive SES. They predicted magnitude M from L, the length of dipole that received SES, and ΔV, the change in Volt or amplitude of the SES (although VAN did not explicitly mention it) using a formula of $(0.32\sim0.37)\times \log(\Delta V/L)$.

Electrochemical, magnetotelluric and the use of electric devices all affected SES. For example, electric train was reported to produce a cultural interference in 15km. VAN missed an earthquake of M5.8 at (39.3, 23.6) on Mar.19, 1989, and made a false prediction on Apr. 3, 1988. They predicted that

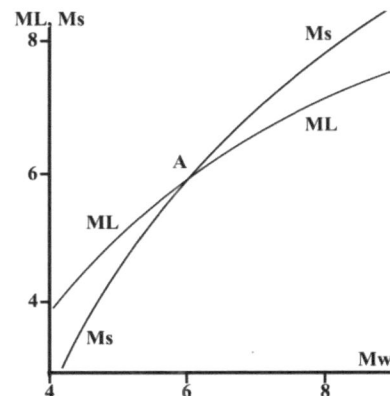

Fig.48 A comparison between ML and Ms
Two curves reveal relationships of moment magnitude (Mw) vs. both local (Richter) magnitude ML and surface-wave magnitude Ms. They have a crosspoint at A (about 6 of ML and Ms). This chart refers to Magnitude Scale and Quantification of Earthquakes (Kanamori, 1983) and Earthquake Magnitude Scales (McCalpin, 1996).

an M5.0 would hit a place 100km north of Athens within 11 days or no later than 20 days after Apr.3. However, they could find only an M3.8 quake 144 km East and Northeast of Athens. Using a formula Ms=ML+0.5, VAN claimed a successful ratio of 15/17 for the predictions from May 15, 1988 to Jul.23, 1989 (Varotsos and Lazaridou, 1991).

In 1996, VAN sent Geophysical Research Letters (GRL) a paper: "Basic principles for evaluating an earthquake method". Editor Geller (1996) arranged a debate between supporters and critics with two catalogs to compare predictions to (Table 7).

Wyss (1996) criticized VAN's using incorrect transformation and denied VAN's success. Wyss also pointed out that 25% of events were without "evidence that were not explosions or other cultural disturbances." Wyss and Allmann (1996) criticized VAN for not offering clear hypothesis on the mechanistic relation between electrical signals and earthquakes, and no precise definitions of parameters in their predictions.

The above criticisms are valid. Fig.48 illustrates the reason why VAN's transformation is incorrect. A prediction should have three closed intervals of time, area and magnitude. It is not valid to predict an earthquake at a point, and then freely claim the prediction success if the epicenter is somewhere else. A major problem for judging VAN predictions is that the inconsistency between the two earthquake catalogs demonstrates their poor quality, and leaves any conclusions based on them questionable.

SES or seismic electric signals do not have a hypothesis of how an earthquake causes it. The Earthquake Vapor Model will attempt to do. Friction among vapor, water, and rock of an impending hypocenter produces not only heat, but also electricity. The electricity forms an electric field. When its electric potential energy reaches certain degree, SES may form from the origin to either the atmosphere, or the ground. In the former, a lightning forms, and this will cause VAN method to miss the precursor. In the latter, a miss is also possible when stations are not placed densely enough. As an empirical method, it is difficult to set up a relationship between the amplitude of SES and magnitude of an earthquake. Many factors such as electric field strength, whether the electric field is dispersed or focused, resistance between the field and the detector, humidity, and so on may affect the amplitude of SES. As an empirical method, it is difficult to find the longest duration from a precursor to its following earthquake. Shou spent 17 years and used thousands events to find 112 days as a possible longest interval between vapor eruption and earthquake. The 11-day time window in VAN (Varotsos and Lazaridou, 1991) may be too small.

Table 13 Chinese Electromagnetic Predictions and misses

a Date	Latitude	Longitude	Ms	Location
20060705~12	41.5~45.5	10.5~12.5	5.2	Italy
	-6	29	5.7	Congo-Tanzania
	1~2.5	96~99	5.5	Sumatra-Nias
	-11~-7	120~124	5.7	Flores-Timor
	58~63	-155~-151	5.5	Alaska
	49~51	-130~-127	5.5	Vancouver
20060719~26	3~8	-34~-31	5.5	Atlantic
	10~14	92~95	5.7	Andaman IS
	44~46	97.5~102.5	5.0	Mongolia
	25~27	99~102	4.5	Yunnan
	17~19	145~148	5.2	Mariana
	43~47	147~154	5.7	Kuril
	51~53	-173~-166	5.8	Alaska
	51~53	176~179	5.8	Aleutian
	51~56	151~156	5.8	Russia
	3	135	5.5	Papua
	-5	148	5.5	PNG

b Date	Latitude	Longitude	M	Location
20060705	23.74	122.21	4.0	Taiwan
20060710	48.17	117.30	4.2	Inner Mongolia
20060715	24.05	101.16	4.7	Yunnan
20060717	32.98	96.21	4.9	Qinghai
20060718	32.78	77.38	4.2	Xizang
20060719	32.97	96.21	5.2	Qinghai

Note: (a)The Chinese group in Xian sent Shou two sheets of predictions by electromagnetic: one at 15:40 (UTC 22:40) on Jul.5, 2006 and the other at 16:10 (UTC 23:10) on Jul.19 with a time window of a week for a total of 17 earthquakes. However, none is correct.

(b) Six earthquakes hit China on Jul.5~19. None of them was predicted. The predictions were witnessed by Chen, I-wan, advisor of the Committee of National Hazard Prediction of China Geophysics Society. Earthquake data were from the USGS.

3.10.11 Chinese Predictions based on Electromagnetic precursor

Campbell (1998) criticized the United Nations for abusing the public fund to support Chinese scientists on the geomagnetic precursor. It had a time window of 1~22 months, an area window of 550km, and a rate of 82.7% for false predictions. The Chinese concluded five causes for high false predictions: "(1) Instrument troubles, such as damp instruments, changed scale values, etc; (2) Changed environment around the station, such as the building of factories and houses near the station: (3) Data collected were too scanty, so that our understanding of the normal background and magnetic anomalies related earthquake cases were not abundant; (4) The magnetic anomalies are correspondent to the weather disasters rarely happening in a hundred years (big floods, droughts, high temperatures, freezes, etc.); (5) Phenomenon which we have not yet recognized."

From Jul.5 to 26, 2006, the Chinese sent Shou two sheets of predictions, together 17 earthquakes based on electromagnetic (Table13a), but none was correct. On the other hand, six moderate earthquakes hit China on Jul.5~19, 2006, but none was predicted (Table13b). The criticism of Campbell (1998) is valid, but his claim of "Deterministic short-term prediction is not possible at present" is wrong.

3.10.12 Seismic Electromagnetic

Seismic Electromagnetic Wave (SEW) does not have a hypothesis to explain how an impending earthquake induces SEW, and a character to distinguish SEW from other electromagnetic waves (Karakelian et al., 2000; Sadowski, 1982). Sadowski (1982) reported that Cherniyaphski observed a sudden increase of 59V from 16:00 to 16:18 in a calm weather in 1924 with an electrometer and then the meter surpassed the largest calibration (400V) in 5 seconds (In general, this needs 15 seconds). At 20:23, an M4.5 happened. Cherniyaphski further adopted a double wire electrometer to measure atmosphere electric potential with artificial explosions underground and got important results: (1) an explosion accompanies an obvious pulse increase of electric potential; (2) the sudden increase of electric potential happened with the sound of an explosion simultaneously, but fraction of a second earlier than the ground tremor; (3) the stronger an explosion and the closer an observatory, the bigger increase an electric potential. A big difference between an explosion and an earthquake is the time span from an increase of electric potential to the two events.

The coincidence between the sudden increase of electric potential and the sound of an explosion suggests friction between explosive gas and surrounding air inducing the increase of electric potential. Like explosive gas, earthquake vapor appears suddenly with high pressure, so its friction with surrounding air may induce an increase of atmosphere electric potential, too. Because vapor current is inconstant: small in the beginning and the end, and big in the middle, and wind usually changes its direction and speed, friction between the vapor current and its surrounding air is changeable. The variable friction can cause a variable electric field. The variable electric field can cause a variable magnetic field. The variable magnetic field can induce a new electric field. The above process continues alternately, so a SEW may form.

Relying on the above hypothesis, we can understand why the voltage of the atmosphere increased 59V and surpass 400V suddenly and before an earthquake. SEW comes from friction between the atmosphere and erupting vapor before earthquake. This may be why Sadowski (1982) did not record SEW during earthquake. Relying on the above hypothesis and that earthquake vapor can appear up to 112 days in advance to earthquake, we can also understand why the Chinese 7-day window (Table 13a) is incorrect. Like SES, SEW is costly (Karakelian et al., 2000).

3.10.13 Remote power alignment

Some scientists have argued that an alignment of the Sun, the Moon and the Earth can sometimes trigger an earthquake. They mentioned many big earthquakes on or within three days before or after a new moon or a full moon to support their hypothesis. However, their view is flawed. During a metonic cycle of 19 years, which is a common multiple of the solar year and the lunar month, there are 235 lunar months and 6940 days, so the average of a lunar month was 29.53 days (=6940/235). When one predicts a big quake 3 days before and after new moon or full moon, the total span of 14 days per month covers **47.4%** (=14/29.53) of the time. Among 160 independent big earthquakes in Southern California from 1709 to the end of 1999, 74 (**46.3%**=74/160) quakes occurred in the 14 days spanning new moon or full moon (Table 14). Thus, using new moon or full moon to predict earthquakes does no better than a random guess.

Some claimed an alignment of the nine planets of the solar system can trigger an earthquake. For example, an Indian Ph.D. adopted an alignment of all Planets to predict an earthquake of magnitude 6~6.7 around (30.5, 79.5) with a radius of 30 km on May 3~5, 2005 to a Chinese nature hazard advisor and a producer of the Indian Miditech.

The both men asked Shou about the Indian prediction. Shou replied "No." The reason is that the force exerted on the Earth by the eight planets in a line was very small. Suppose that the masses of all eight plants were concentrated on Mars, the planet closest to the Earth, the force was still only 0.106 of the force between the Moon and the Earth (=$2662 \times 0.384^2/(0.0734 \times 225^2)$), Here: 2662 and 0.0734 are the masses of the eight planets and the Moon in a unit of 10^{24} kg respectively, 225 and 0.384 are the average distances between the Earth and respectively the Mars and the Moon in a unit of 10^6km). Since the Moon can't trigger an earthquake, neither can the alignment of planets.

Indeed, the Indian Ph.D.'s prediction was wrong. There was neither an earthquake of magnitude ≥6 in the East Hemisphere (-90~90, and 0~180) from Apr.29 to May 9, 2005, nor an earthquake of magnitude ≥4 in the area of ±5° (25.1~35.9, 74.1~84.9) around his predicted area from Apr.19 to May 18, 2005. The closest earthquake struck on Apr.18, with a magnitude of 4.1 at (32.7, 76.4). It was far smaller than, and 349 km and 15 days off the prediction. Shou has posted 'Neither the Moon, nor the Planets Can Predict Earthquakes' on May 18, 2005@5.

3.10.14 Animal Behavior

The Haicheng Earthquake Study Delegation (1977) reported, "From December 1974 until the time of the earthquake. …snakes being found frozen on the road and rats being agitated to the point of acting dazed. Those examples were from Tantung, 150km from both the epicenter of Liaoyang swarm and the Haicheng epicenter. After the Liaoyang earthquake, reports of anomalous animal behavior spread …geese flying, chickens refusing to enter their coop, pigs rooting at their fence,…We are unable to assess the significance of those reports in the decision to issue the imminent prediction of the earthquake."

Jinchang and Zhang (1984) reported that events of hibernating snakes crawling out from their holes had two situations: one because of earthquakes, and the other because of local warm air.

This phenomenon appeared before the M7.3 Haicheng earthquake on Feb.4, 1975, the M7.2 Romania earthquake on Mar.4, 1977 and the 6.8 Russia (Kyrgyzstan) earthquake on Nov.1, 1978. This phenomenon also appeared in Linfen, Shanxi in Jan.~Feb, 1976, but an earthquake did not follow. Namely, Animal Behavior has false warning.

The Northridge earthquake and the Bam earthquake were not preceded by abnormal animal behavior. When Shou lived in Pasadena, Southern California, he himself saw a big dog of his neighbor that was calm before the Northridge earthquake. He predicted the big Xinjiang, China earthquakes on Apr.5~6, 1997 to the USGS on Mar.6 (No.17 in Table 11). The earthquakes injured 23 people and killed 100 head of livestock. Abnormal animal behavior was not reported. Abnormal animal behavior can be caused by warm vapors erupting before earthquakes, and may predict earthquakes sometimes. However, abnormal animal behavior can give false predictions and miss earthquakes. Moreover, it is hard to decide when, where and how big an earthquake will be.

3.10.15 Predictions using earthquake clouds by other people

Li (1982) reported that ancient Chinese and Italians had studied special clouds which had been indicative of impending earthquakes. The Chronicle of Lon-De County (35.7, 106.1) in Ningxia province, China, 300 years ago (recompiled in 1935) recorded, "It was sunny and warm; the sky was blue and clear. Suddenly, there appeared threads of a black cloud spanning the sky like a long snake. The cloud stayed for a long time, so there would be an earthquake".

The method was revived in Japan and China. On the morning of Mar.6, 1978 Kagida, the former mayor of Nara city, Japan, predicted the M7.8 Kantow earthquake on Mar.7 by the clouds. He also proposed that the epicenter of an earthquake would be located in the mid-perpendicular plane of the clouds (Li, 1982), which later proved to be not general (Shou 1999). Following this successful prediction, there was a brief period of activity in the scientific communities in China and Japan.

Unfortunately, those people did not have a model to explain why earthquake clouds could predict earthquakes, nor how to predict earthquakes by the cloud. They supposed earthquakes to happen one or two days after the cloud, which is wrong (Shou, 2006b).

3.11 Common problems among precursors other than earthquake vapor

Mormile (1994) reported that Japan spent $100 million a year for several years to build near Tokyo an integrated network of various sensors, such as seismic activity, rock strain, crustal tilt, and tide and ground-water level. However, this network missed the M6.9 Kobe quake on Jan.16, 1995. Varotsos and Lazaridou (1991) reported that Greece established a network of 17 SES stations. However, it missed the M6 Athens quake on Sep.7, 1999. Heaton and Wald (1994) reported that the USGS had monitored Los Angeles by the best means of surveillance, but missed the M6.7 Northridge quake on Jan.17, 1994. On the other hand, the USGS made two Parkfield predictions and one Los Angeles prediction, but nothing happened. The Chinese predictions by electromagnetic precursor all failed, while the quakes in China were all missed (Table 13). What is the common problem?

Geller et al. (1997) concluded, "There are no objective definitions of 'anomalies', no quantitative physical mechanism links the alleged precursors to earthquakes, statistical evidence for a correlation is lacking, and natural or artificial causes unrelated to earthquakes have not been compellingly excluded." These reasons all get points. For example, Rn does not have a mechanism; foreshock and aftershock even do not have scientific definition.

However, Geller et al. (1997) missed the most important precursor: earthquake vapor and its related phenomena. These phenomena include sudden appearance, abnormal temperature, high pressure, and fixed nozzles of vapor eruption, as we have discussed in Chapter 1. Abnormal temperature has a scientific definition (Shou, 2011), The Earthquake Vapor Model has a characteristic curve of "Dehydration" relying on experimental data (Fig.4). It can explain why vapor eruption can trigger an earthquake (Harrington and Shou, 2005). All other precursors lack such a curve to explain why earthquakes happen after those alleged precursors.

All nine devastating earthquakes of death toll ≥10,000 in the World from Jan.1, 1990 to Dec.31, 2015 (Table 8) were preceded by earthquake vapor (Fig.6, 13e~k. 17b~d, 30a, 37~41). The four linear Sichuan clouds in 2008 predicted four big earthquakes (Fig.39) which were the only big ones in an area of 4.2 million km^2 (20~40 and 100~120) and a period of more than five years (Jun.3, 2007~Apr.19, 2013). This demonstrates that the earthquake vapors can be used as a reliable precursor to make short-term predictions.

The Earthquake Vapor Model has successfully generated the Bam prediction (Fig.7a) and the Hector Mine prediction (Fig.32b); the set of Shou's 63 independent predictions (Table 11) demonstrated that the model far exceeded the performance expected from random guesses ("Values of the Vapor Precursor" **3.9**).

Shou has predicted these earthquakes only by eyes, camera, compass, computer, paper, pen, satellite image and the Internet without any fund, whereas other precursors cost a lot. The above problems make various precursors unsuccessful (Geller et al. 1997).

Chapter 4 Critiques of the Plate Theory

4.1 History of the Plate Theory

The Plate Tectonics Theory attempts to explain phenomena on geography, paleoclimate, paleobiology and seismology. The USGS explains, "A *plate* is a large, rigid slab of solid rock. The word *tectonics* comes from the Greek root 'to build'." 'Plate tectonics' refers to "how the Earth's surface is built of plates". "The *theory of plate tectonics* states that the Earth's outermost layer is fragmented into a dozen or more large and small plates that are moving relative to one another as they ride atop hotter, more mobile material" @22.

In 1596, Abraham Ortelius, Dutch mapmaker, first proposed the Americas being "torn away from Europe and Africa . . . by earthquakes and floods" in his *"Thesaurus Geographicus"*. In 1858, French mapmaker Antonio Snider published maps depicting continents adrift.

In 1912, Wegener (2002 trans.) German meteorologist independently published "The origins of continents". He contended that all continents had once jointed into a super continent, called "Gondwanaland", which began to split apart around 200 million years ago, and drift to current locations gradually. His theory attempted to explain puzzles of how coastal geomorphic features of the Americas and Europe-Africa could be extremely similar, how fossils of tropical flora and fauna could be dug up from the Arctic (e.g. Northern Greenland), how moraines on abraded basal surfaces could be found in a vast area of "Australia, South Africa, South America and above all in east India", and so on. The Wikipedia@23 comments, "Wegener failed to resolve: what is the nature of the forces the plates?" and mentions, "Arthur Holmes, British geologist, proposed in 1931 that the Earth's mantle contained convection cells that dissipated radioactive heat and moved the crust at the surface" (Holmes, 1931).

Hess (1962) adopted "Convection Cells" and proposed "Seafloor Spread" to explain "Continental Drift". Vine (1966) proposed "Magnetic anomalies" in oceans to support "Seafloor Spread". Le Pichon (1968) and Isacks et al., (1968) replaced Wegener's coast line by earthquake epicenter for boundary of plates. In 1989, the Department of Interior and the USGS published a world map of volcanos, earthquakes and plate tectonics, titled: "This Dynamic Planet" (Fig.49). However, many plates do not have complete boundaries, e.g. N_1 between the Eurasia and the North America, N_2 between North America and South America, etc. (Fig.49).

Not all scientists advocate the plate theory. For example, Vening Meinesz, F.A. (1887-1966), famous Dutch geodesist and president of the International Union of Geodesy and Geophysics, thought that the mobility of the Earth's crust was local and continental drift was impossible@24. Beloussov (1907-1990), head of the Geodynamics Department of the Institute of the Physics of

@22 http://pubs.usgs.gov/gip/dynamic/historical.html#anchor9588978
@23 http://en.wikipedia.org/wiki/Continental_drift
@24 http://www.egu.eu/egs/meinesz.htm

the Earth in Moscow, published "Against the hypothesis of ocean-floor spreading" in 1970@25. Professor Meyerhoff, A.A. (1928-1994) and his father Professor Meyerhoff, H.A. (1899-1982), chief hearing officer of the National War Labor Board under Presidents: Roosevelt and later Truman, published, "The New Global Tectonics: Major Inconsistencies"@26. Meyerhoff (1972) described the debate between them and the main stream scientists as "The blind men and the elephant".

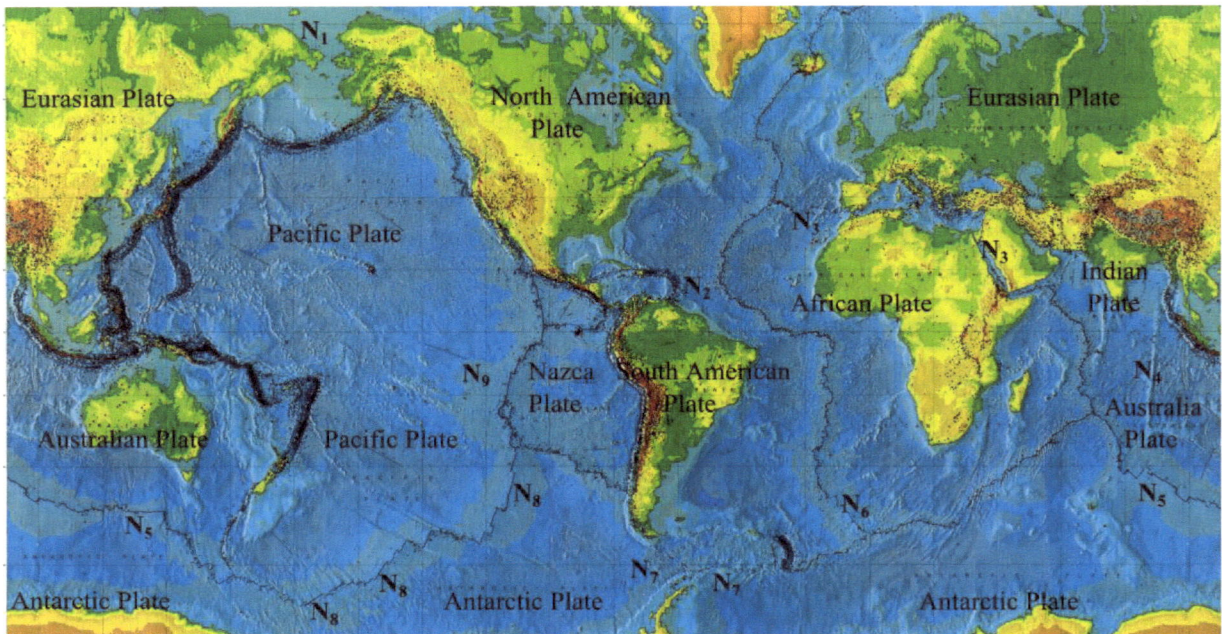

Fig.49 This Dynamic Planet

N_1 reveals no boundary between the Eurasia Plate and the North America Plate, and so do N_2 through N_9. This "World map of volcanoes, earthquakes, impact craters, and plate tectonics" was compiled by Simkin, T; Unger, J. D; Tilling, R. I; Vogt, P. R. and Spall, H with a title "This Dynamic Planet", published by the U.S. Department of the Interior and the U.S. Geological Survey in 1994.

4.2 Problems of Continental Drift Hypothesis
4.2.1 Americas drift

"The strongest argument for continental drift" was "the separation of South America from Africa since the end of the Paleozoic" (Hess, 1962). Hough (2002) amazed, "No schoolchild who looks at a globe can help but be struck by an observation as simple as it is obvious: South America and Africa would fit together if they could be slid magically toward each other. Surely this is not a coincidence; surely these two continents were at one time joined."

Relying on computer technology, South America can be simulated to slide toward Africa on map. Fig.50a~b shows this attempt. The brown and blue outlines are South America coastline slid to and turned counterclockwise $28°$ and $44°$ respectively to best match Africa. However, both

@25 http://en.wikipedia.org/wiki/Vladimir_Belousov
@26 http://archives.datapages.com/data/bulletns/1971-73/data/pg/0056/0002/0250/0269.htm

outlines overlapped with Africa over a common area of about 75,000 km^2 (**de** Fig.50b) and thus do not completely match Africa. Moreover, continents are three-dimensional. Continental shelves, which are underwater landmasses, extend beyond continental coast circles. Fig.50a shows part of continental shelf outline (l through q) extending beyond coast lines (a through h). Therefore, even if their coast lines matched each other completely, the continents might not have jointed previously.

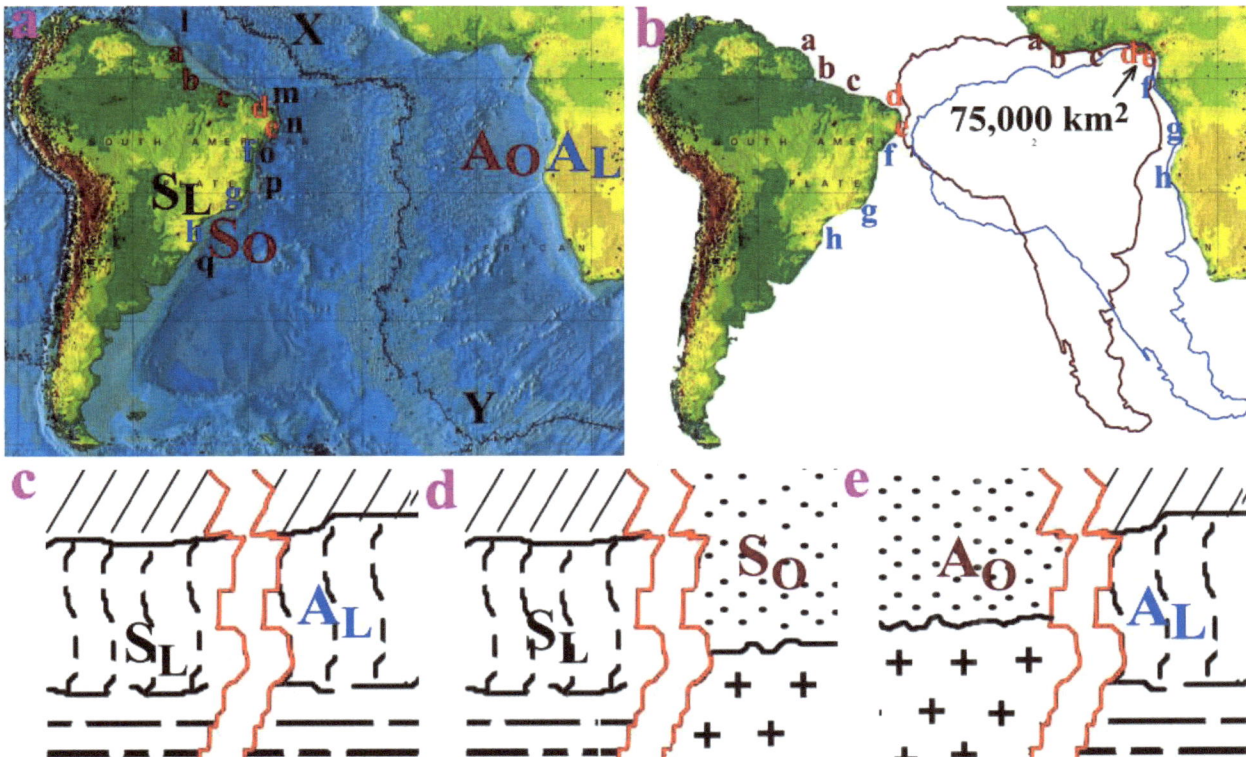

Fig.50 Problems of the hypothesis that South America drifted from Africa
(**a**) Letters a (brown) through h (blue) depict part of the east-coast of South America. S_L in the South America and A_L in the Africa are supposed to be a pair of once jointed points for Wegener's plates. S_O and A_O in the Atlantic respectively present the ocean vicinity of S_L and A_L. Curve XY in the Atlantic is part of "Mid-ocean ridge" and new boundary between the two plates by the "New plate tectonics". Brown curve tagged by letters l through q (black) is assumed as part of the shelf of the South America.
(**b**) Brown and blue outlines represent the coast line of the South America except moved and turned 28° and 44° counterclockwise respectively to best match the coast line of the Africa. The brown outline matches a part of the coast marked by letters a through d; while the blue by letters f through h. However, both outlines overlap the Africa with an area between points d and e of about 75,000 km^2 (red) and do not match Africa.
(**c-e**) If the continental drift hypothesis were true, then S_L of the South America plate and A_L of the Africa plate would show similar strata at the once jointed points (**c**). In contrast, S_L and its Atlantic vicinity S_O would show distinct strata (**d**), and so would A_L and its Atlantic vicinity A_O (**e**). The map was from U.S. Department of the Interior and the USGS.

Wegener (2002 trans.) claimed his most convincing reason to support the continental drift hypothesis to be a "continuation" of the Carboniferous coal deposits from Europe to the Atlantic Ocean, and then to North America. His claim of "continuation" contradicts his claim of "separation" between Europe and North America. The necessary and sufficient evidence for once jointed continents is both similar strata along the separated sections (Fig.50c) and different strata between the separated section of a continent and its vicinity in the ocean (Fig.50d~e). However, no effort has been invested in comparing the strata structures of continents and their ocean vicinity. Moreover, if Americas had separated and drifted from Europe-Africa, earthquakes and volcanoes should have happened along the separated coasts. However, the recorded epicenters and volcanoes in Fig.49 are not consistent with this.

4.2.2 Permian or Carboniferous glaciation

Wegener (2002 trans.) claimed, "One of the strongest proofs of these ideas are to be found in Permian glaciation (some say Carboniferous), the traces of which have been observed at some places in the southern hemisphere, but are missing in the northern hemisphere. ... These undoubted moraines on abraded basal surfaces are found in Australia, South Africa, South America and above all in east India." Wegener thought, "Such a large extent of a polar ice cap is impossible", and concluded those continents once jointed. Since Permian (-300~-250Ma, i.e. 300~250 million years ago) and Carboniferous (-360~-300Ma) both lasted long periods of time, it is possible that moraines in Australia, South Africa, South America and east India had happened in distinct periods of time (say, each lasting one million year out of the 50 or so million years), and may have never been physically connected.

4.2.3 Polar wander

One piece of "evidence" for continental drift is the polar wander phenomenon. Wegener (2002 trans.) proposed the "polar wander" hypothesis to explain why fossils of tropical and subtropical plants in Tertiary (-65~-2Ma) existed in "Greenland Grinnell land, Barren Island, Spitzbergen – locations that are currently 10~22° north of the tree line." To support this, he cited Nathorst's work which claimed that the North Pole was at 70°N and 120°E during Europe glaciation. Thus, Kamchatka, the Amur lands, and Sakalin were at latitude of 67–68°N, while Spitzbergen, Grinell land, Greenland were at 64, 62, and 53–51°N respectively instead of their current locations at 79, 80, and 60~80°N. This would allow him to account for the fossils in Northern locations such as Greenland. However, Wegener could not explain why the coeval South Pole in South Africa did not have glaciated fossils.

We provide an alternative hypothesis that could resolve the difficulty encountered by the Polar Wander hypothesis. We have discussed that earthquake vapors and earthquakes release enormous heat. If earthquakes were frequent, the warm temperature conditions might support tree growth even in very Northern locations. Fig.51 shows a coincidence between the locations discussed by Wegener and locations of earthquakes of M ≥5 in latitude ≥40N from 1990 to 2012. This suggests that those places had a long history of earthquakes, and that fossils of tropical and subtropical plants in Greenland reflected heat generated from earthquakes. Thus, it is unnecessary to evoke "Polar Wander" and its problem of no glaciated fossils in the coeval South Pole.

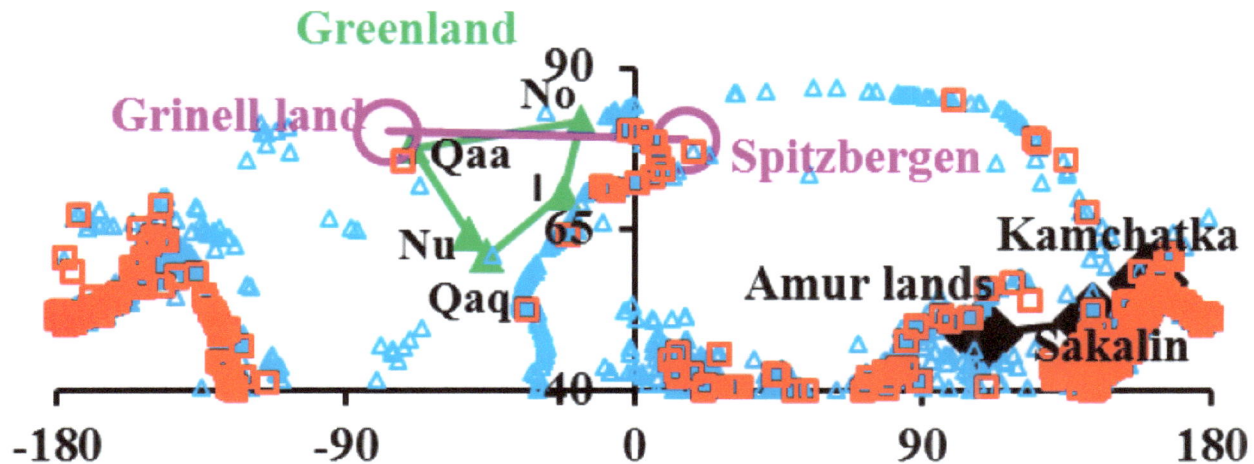

Fig.51 Modern Earthquakes vs. Ancient Greenland Warming
Cyan triangles and red squares respectively plot earthquakes of magnitude 5~5.9 and ≥6 in latitude ≥40 from Jan.1, 1990 to Dec.31, 2012. Green solid triangles plot five Greenland cities: No (Nord: 81.6, -16.7), Qaa (Qaanaaq: 77.5, -69.2), Nu (Nuuk: 64.2,-51.7), Qaq (Qaqortoq: 60.7, -46) and I (Ittoqqortoormiit: 70.5, -22). Magenta circles plot Grinell land (80, -77) and Spitzbergen Island (78.8, 16) near Greenland in "10~22° north of the tree line" (the tree line would then be *roughly* 60°N). Black solid squares plot Kamchatka Peninsula (57, 160), Sakalin Island (51, 143), and Amur lands (48.8, 108.8), places where temperature changes have been hypothesized to be caused by polar wander. Earthquake data and coordinates of above cities and lands are from the USGS and Google search respectively.

4.2.4 Distance increase as evidence for continental drift?

Wegener (2002 trans.) cited the increasing distance between Cambridge and Greenwich over time as an evidence for continental drift. "According to Schott, the three great measurements of length from 1866, 1870 and 1892, show the following values of distance (time) differences between Cambridge and Greenwich: 1866: 4 h, 44 m, 30.89 s; 1870: 4 h, 44 m 31.065 s; 1892: 4 h, 44 m 31.12 s." This measurement through the trans-Atlantic cable was regarded as "a much more exact determination" of Continental Drift. However, a drooping cable between two fixed poles can increase its length due to material stretching by weight or current for example.

4.3 Problems of Wegener's disciples' works
4.3.1 Hess' Seafloor Spread

Hess (1962) pointed out, "Mid-ocean ridges have high heat flow, and many of them have median rifts and show lower seismic velocities than do the common oceanic areas. They are interpreted as representing the rising limbs of mantle-convection cells. The topographic elevation is related to thermal expansion, and the lower seismic velocities both to higher than normal temperatures and microfracturing. Convective flow comes right through to the surface, and the oceanic crust is formed by hydration of mantle material starting at a level 5 km below the sea floor. The water to produce serpentine of the oceanic crust comes from the mantle at a rate consistent with a gradual evolution of ocean water over 4 aeons" (1 aeon $=10^9$ years).

"The Mid-Atlantic Ridge is truly median because each side of the convecting cell is moving away from the crest at the same velocity, ca. 1 cm/yr. A more acceptable mechanism is derived for continental drift whereby continents ride passively on convecting mantle instead of having to plow through oceanic crust".

This hypothesis seems rational, but has serious problems due to the numerous (nineteen) assumptions. The more assumptions, the more likely conclusions are wrong. For example, the continental separation was supposed to be evidence. Hess (1962) mentioned, "The old evidence, which was the strongest argument for continental drift, namely the separation of South America from Africa since the end of the Paleozoic". However, our discussion by Fig.50 shows that postulating continental separation is problematic.

Another problem is self-conflict. Hess (1962) divided the seafloor into four layers: two sediments of a total thickness of 1.3km, a serpentine of 4.7km, and a peridotite of 29km, together 35km with water about 5km above. He emphasized that the thickness of layer 3 was "striking" for more than 80% seismic profiles showing it to be 4.7±0.7km (Fig.52a,b) with an average velocity of 6.7km/s (Fig.52b). Namely, this striking thickness was not obtained by wide drilling, so its reliability needs to be verified. Moreover, serpentine (Layer 3) and peridotite (Layer 4) do not have a difference on chemical composition. Layer 3 is just 70% hydrated or serpentinized peridotite. If one measures seismic velocity or hydration from the top of Layer 3 to the bottom of Layer 4 gradually, the linear character of *Fig.3* of Hess (1962) will deny a boundary between Layer 3 and Layer 4 or the "striking" thickness.

Another self-conflict hides in the two columns of Fig.52b. Hess (1962) noticed "Seismic velocity decreased by higher temperature and fracturing" (Fig.52b), but its data in each column showed that the lower the elevation, the higher the both temperature and seismic velocity (Fig.52b). Namely, the former indicates higher temperature corresponding to lower seismic velocity; while the latter does opposite.

The hypothesis has conflicts. For example, mantle flows in Layer3 (Fig.52a,b), 1.3km under seafloor or 6.3km under sea level. However, drilled wells, whose deepest reaches 12.376 km, produce oil about 4×10^7 m³/d in offshore of the Okhotsk Sea@27. Oil comes from ancient sea animal, so the depth of 6.3km indicates ancient sea animal living in mantle. This is not true. Moreover, many large earthquakes are deeper than 40km. They usually cause large gaps, by which, volcanos should erupt. However, this is not true, either. Maybe, because it is too hard to explain earthquake in Moho and because the deepest quake has a depth of about 700km, Le Pichon (1968) proposed the "lithosphere" to have a "depth as great as 700 km". The above problems question the highest elevation of the mantle flow.

The convective cell model (Fig.52c) lacks evidence. According to this model, cooler descending limbs of the West Atlantic Cell and the East Pacific Cell should be paired with descending limbs of the Americas Cell. Both descending limbs of the America Cell should come from a common warmer rising limb like the red arrow of Fig.52c, which should form ridges as

@27 http://en.wikipedia.org/wiki/Sakhalin-I

- 90 -

those found in Atlantic and Pacific. However, Hess did not show any such evidence. Moreover, He claimed that descending limbs released water (about 500°C) to oceans, but he did not show such evidence either.

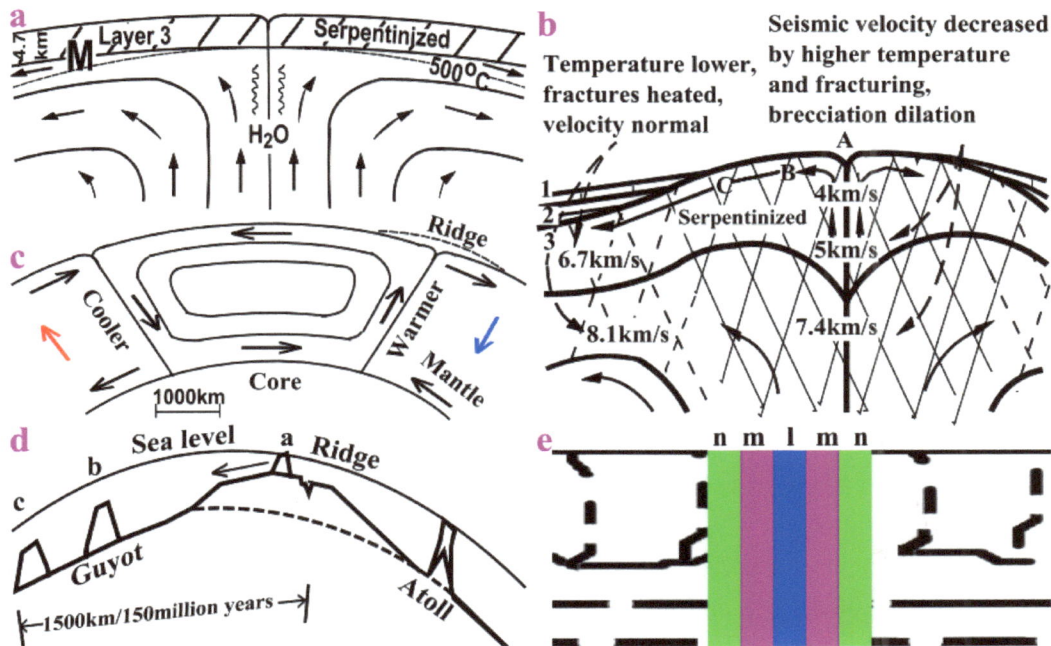

Fig.52 Problems of Seafloor Spread

(**a**) Hess' schematic diagram of formation of even thickness serpentine.

(**b**) Diagram to represent progressive overlap of ocean sediments (Layers 1 and 2) on a mid-ocean ridge (below sediments) and the postulated fracturing (thin crisscrossing lines) where convective flow (arrow) changes direction from vertical (toward point A) to horizontal (point B to point C). Fracturing and high temperature are used to account for the lower seismic velocities on ridge crests (point A), while cooling and healing (dotted crisscrossing lines) of the fractures are used to account for normal velocities on the flanks.

(**c**) Hess' diagram of a mantle convection cell. Red arrow and blue arrow represent a deduced rising limb and a descending limb respectively.

(**d**) Volcanic peaks (point a), guyots (points b and c), and atolls are postulated to migrate from the crest of a ridge to the flanks with a rate of 1cm/yr or 1500km/150million years.

(**e**) P and Q: initial strata. If color bands l, m and n represent current to past mantle eruptions from the mantle core to the seafloor covering an entire ocean ridge, then this would truly be "Seafloor Spread".

Figures **a**~ **d** are simulated diagrams from Hess (1962).

The hypothesis that volcano drift represents seafloor spread lacks logic. Hess (1962) depicted, "Volcanoes truncated on the ridge crest move away from the ridge axis at a rate of 1 cm/yr. Eventually they move down the ridge flank and become guyots or atolls rising from the deep-sea floor" on the old Mesozoic mid-Pacific Ridge (Fig.52d). However, this model was for truncated volcanoes standing alone with wide free room, not for plates squeezing against each other.

A logical hypothesis of seafloor spread has two requirements. The first requirement is a section like that in Fig.52e. Vertical bands l, m, and n etc represent fissures or vents from the core to the seafloor. Repeated eruption (first n, then m, then l etc) will ensure that the fissures or the vents themselves spread. The other requirement is a continuation of this section throughout vicinities of the fissures or the vents to ensure whole ridge spread. By such a hypothesis, it is easy to divide between true spread and false spread. For example, Hess' (1962) claim: "The oceanic crust … starting at a level 5 km below the sea floor" does not work because Layer 4 does not spread and there was no discussion on what happens to the vicinities.

We have discussed main problems. Now, let's use the Earthquake Vapor Model to account for the phenomena Hess discussed. Earthquakes exist in the Mid-ocean ridges (Fig.49), therefore rifts, water penetration, heat flow, eruption of earthquake vapor, rock and sand, and seismic activity are normal there. Earthquakes occur more frequently in shallow regions than in deep regions (Fig.1), and so are rifts and hydrations. Thus, earthquake wave propagates slower in shallower regions because as suggested by *Fig.3* of Hess (1962) the lower rock density (caused by more hydrated fissures) in shallower regions cause slower seismic velocity. This is consistent with measurements in Fig.52b.

4.3.2 Vine-Matthews-Morley's magnetic stripes

Magnetic stripes are admired as "The true Rosetta stone of plate tectonics — the key that won over a skeptical community", and "simpler than hieroglyphics". The stripes are described as "The generation of magma at mid-ocean ridges during alternating period of normal and reversed magnetism" (Hough 2002). Fig.53a~c is their well-known diagram. This hypothesis is named after Vine-Matthews-Morley. Vine and Matthews (1963) claimed the phenomenon of magnetic stripes as a "corollary" of seafloor spread.

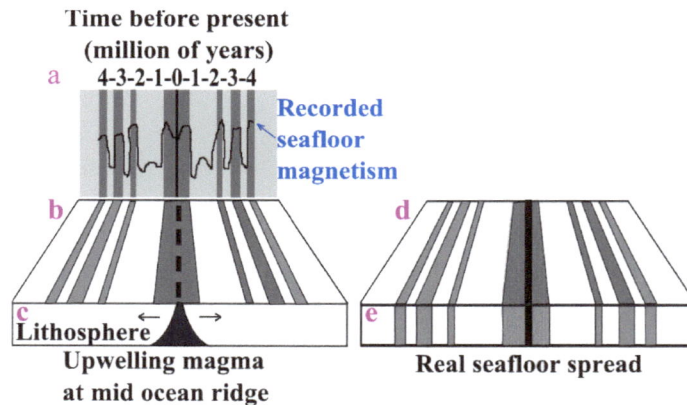

Fig.53 Problems of Magnetic stripes
(**a-c**) This schematic diagram of the formation of hypothesized seafloor magnetic stripes is adapted from Wikepedia@28 and Hough (2002). Dark and white bands correspond to magnetic alignments: parallel and antiparallel to the present day field direction respectively. The recorded seafloor magnetism (Black curve in "**a**") reflects the alternate orientations. Black dashes in "**b**" may represent nozzles of volcanos, fissure eruptions, or lave flows. A pyramid in "**c**" indicates upwelling magma, and two arrows depict seafloor spread.
(**d-e**) If seafloor spread hypothesis were correct, the eruption nozzles would connect together as "**d**" and the stripes would penetrate into the core to separate the seafloor as "**e**".
@28http://en.wikipedia.org/wiki/Vine–Matthews–Morley_hypothesis

We have discussed that Hess's seafloor spread hypothesis does not work, thus its "corollary" is also flawed. To supply evidence for seafloor spread (which is a large-scale phenomenon), the nozzles have to be connected in a continuous fashion (Fig.53d). Furthermore, since eruption comes from lava layer (Asthenosphere), *all* magnetic stripes need to penetrate to lava layer like in Fig.53e. However, no such evidence exists.

Vine and Matthews' (1963) investigated bathymetry (water depth) and magnetism from the Mid-Atlantic ridge and the Carlsberg ridge, Indian Ocean. They claimed three kinds of anomalies: "(1) long-period anomalies over the exposed or buried foothills of the ridge; (2) shorter-period anomalies over the rugged flanks of the ridge; (3) a pronounced central anomaly associated with the median valley". However, they offered no definition for terms such as "long-period anomalies", "shorter-period anomalies", "pronounced central anomaly", and "median valley". Neither does the figure (*Fig.1* of Vine and Matthews, 1963) attempt to mark or distinguish these three kinds of anomalies.

Even the hypothesized mechanism of the formation of magnetic stripes is confusing. Vine and Matthews (1963) stated, "A trough of negative anomalies … separates two areas of positive anomalies … corresponding to mountains on either side", and "The bottom topography indicates the relief of basic extrusive such as volcanoes and fissure eruptive". The authors did not explicitly state whether mountains with positive anomalies correspond to positive magnetic stripes. If so, the two mountains should be of the same age. However, in the diagram (Fig.53a), the two mountains of positive stripes are of different ages. If not, where are those positive stripes? Moreover, magnetism should not be measured by bathymetry. This is because magnetic strength on the trough should be a function of the thickness of the lava relief in the trough.

To substantiate their reversal magnetic hypothesis, Vine and Matthews (1963) cited "the result from the preliminary Mohole drilling" (Cox and Doell, 1962 and Raff 1963). The Mohole drilling had a depth of 3.568km, located at off the coast of Baja, California, and was claimed to have a reversal magnetic basalt lava flow. The above report gave the lithosphere a thickness of less than 4km. However, even though the true thickness of lithosphere is unknown, lithosphere thickness should be much deeper than 4km. This is because many large earthquakes had a depth near 700km, and earthquakes should not occur in lava flow. Maybe because of this, Le Pichon (1968) proposed the "lithosphere" to have a "depth as great as 700 km". The Mohole drilling conclusion cited by Vine and Matthews conflicts with the Plate Hypothesis, which supposes earthquakes to be in faults between "solid" plates@29. Le Pichon's proposal is likely correct, and should be tested through more work, e.g. rocket drill in different places.

4.3.3 New plate tectonics

By the late 1960s, the continental drift hypothesis had evolved to the "New Plate Tectonic" theory. The plates were assumed to be rigid and have seismic active boundaries along mountains, mid-ocean ridges and major faults, different from the coast boundary proposed by Wegener. For

@29 http://geomaps.wr.usgs.gov/parks/deform/gfaults.html

example, Le Pichon (1968) divided the globe into six rigid plates: Pacific, America, Antarctica, Africa, Eurasia, and India by earthquake active belt. He adopted the mid-Atlantic ridge as the new boundary between Americas and Europe-Africa (Fig.54a). He supposed that the "only modifications" of the plates would "occur along some or all of their boundaries". He proposed five rotation centers for South Pacific, North Pacific, Atlantic, Arctic and India during the Plio-Pleistocene time to support the continental drift. Isacks et al., (1968) admired Le Pichon's model. They claimed that seismology supported the new tectonic theory. However, they wondered about how the new tectonic theory might explain major earthquakes outside the boundaries. The USGS published "This Dynamic Planet" (Fig.49) and defined three kinds of faults: normal, reverse, and strike-slip (Fig.54b) for the new tectonic.

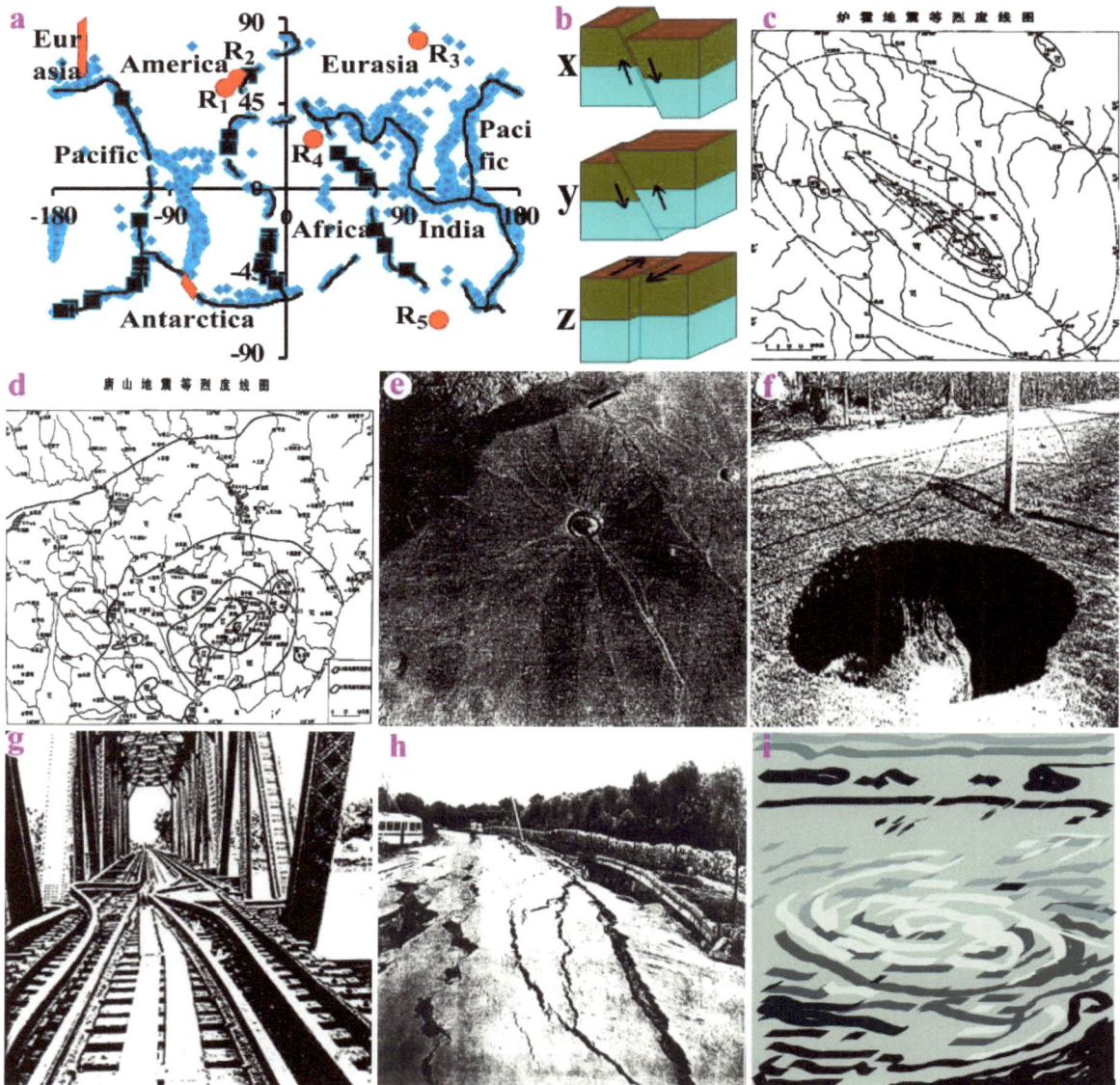

Fig.54 Problems of New Global Tectonics

(a) Le Pichon (1968) divided the globe into six plates by seismic active boundaries (black curves), and the so called "arbitrary" boundaries (red). He claimed five rotation centers R_{1-5} (red circles) respectively for North Pacific, Atlantic, Arctic, India and South Pacific during the Plio-Pleistocene period. Cyan diamonds: all earthquakes ($M \geq 6$) in the World from Jan.1, 1990 to Dec.31, 2012, reported by the USGS.
(b) According to the different types of relative movement (inverse arrows), the USGS divides "faults" into: normal (x), reverse (y), and strike-slip (z) @29. Color brown depicts surface, and green and cyan reveal different layers.
(c-d) Isoseismal maps for the M7.9 Luhuo (31.5, 100.4) earthquake in 1973 and the M7.8 Tangshan earthquake (39.46, 118.21) in 1976 respectively.
(e-h) Photos from the Tangshan earthquake: a spring hole while erupting water and soil, a sinkhole (diameter 3m, depth 3m), a bent railway, and ground crevices. c-h were from the Institute of Geology, State Seismological Bureau (1981).
(i) A drawing for a vortex of the 2011 Japan tsunami. Ref. a photo of Google and Yahoo.

Le Pichon's (1968) hypothesis is confusing. He used earthquake active belts as a mechanism for separating plates, but the belts were not continuous. Thus he used "arbitrary" boundaries to fill the major gaps: one between America and Eurasia, and another between America and Antarctica for separation (red Fig.54a), even though these arbitrary boundaries had no recorded earthquakes (Fig.49). Besides these major gaps, the proposed belts have more other discontinuous regions without recorded earthquakes. This suggests that the plates have not yet separated. Moreover, earthquake active belts do not separate plates because no lava flows out of earthquakes. These reasons refute the "New Plate Tectonic" hypothesis.

The rotational regions associated with the four rotation centers were not described. Le Pichon, (1968) only showed the region of rotation for the India plate (Fig.54a). However, even the Indian rotation only has data from its west edge, which is insufficient. Since the rotation edge is not circular, the rotation of the India plate would clash with neighboring plates, which is unlikely. The Pacific plate is supposed to have two rotation centers: one for the South, and the other for the North. However, if the rotation boundaries had existed, then the number of plates would have to increase. On the other hand, if rotation boundaries had not existed, independent movement would be impossible. The Atlantic is assumed to be assigned to three plates: America, Africa and Eurasia, so it cannot rotate independently of the three plates. Le Pichon also assumed that the Arctic is not a plate, but rather, its rotation would make "modifications" in plates.

The USGS divides the globe into more plates by earthquake-active belts. This plate definition inherits two basic problems: discontinuous boundaries (N_{1-9} Fig.49) and having internal earthquakes, so this plate definition does not work for the "New Plate Tectonic" theory, either. "Fault" defined by the USGS (Fig.54b) does not explain how an earthquake happens. First, two blocks of a fault have same layers and a relative displacement, which suggests "fault" being not a reason of an earthquake, but a consequence. Second, the first earthquake in any location must by definition have happened where no earthquake had happened before. Thus, the concept of fault is not meaningful.

Neither the New Plate Tectonic theory nor the fault model considers water, heat, and pressure, so they fail to account for real earthquake phenomena. For example, the Tangshan earthquake erupted very hot matter which burnt a man (Shi et al., 1980), and erupted water and soil which damaged a ceiling (Fig.2). The Idaho earthquake created a spring of 115 feet in height (Lane and Waag, 1985). The 1999 Taiwan earthquake erupted rocks to form a hole of 4-meter-wide and 40-meter-deep (Huang et al., 2003). The Bam earthquake occurred at the nozzle of the Bam cloud exactly (Fig.7a). Inner earthquakes in China created smooth isoseismal circles (Fig.54c-d). The Tangshan earthquake produced both pressure and suction to form water-soil springs (Fig.54e), and sinkholes (Fig.54f) respectively. It also produced both push and pull to bend railways (Fig.54g), and form long, large crevices (Fig.54h) respectively. The 2011 Japan tsunami earthquake caused vortexes (Fig.54i).

On the other hand, Vening Meinesz, Beloussov, and Meyerhoof father and son are correct. Indeed, "The mobility of the Earth's crust" is local, and "continental drift" is impossible @24, It is right of "Against the hypothesis of ocean-floor spreading"@25, and "The New Global Tectonics: Major Inconsistencies"@26.

4.4 InSAR image and explanation by earthquake vapor

Synthetic aperture radar interferometry (InSAR) is a technique providing maps of the topography and displacement of the Earth's surface. Fig.55a is an InSAR image. Its raw data were from satellite ERS-2 (European Remote Sensing), a Sun-synchronous polar satellite (1995~2011). Relying on Global Position System (GPS), it can produce precise superimposing images with a repeat cycle of 35 days. Radar waves of two overlapping images always have a phase difference because of, for example, different elevations of the satellite. Combining two overlapping images, interferometric stripes form, waves in phase cause constructive interference and waves out of phase cause destructive interference. When a big earthquake happened between the times of two SAR images, their InSAR image could depict what this earthquake did. Fig.55a and Fig.55b respectively reveal what the 1997 Manyi, Xizang earthquake (Peltzer et al., 1999) and the 1999 Hector Mine earthquake did (Peltzer et al., 2001). A full color cycle (blue-red-yellow-blue) represents 50cm change in Fig.55a, and 10cm in Fig.55b.

Peltzer et al. (1999) proposed an asymmetry elastic model to account for asymmetric lobes of the Manyi earthquake (Fig.55a), and attributed its cause to cracks in the crust. However, Peltzer et al. (2001) later stated that the Hector Mine earthquake (Fig.55b) had "a complicated rupture", recognized "by field observers (USGS)", and "the displacement curves on the western and eastern sides of the fault do not depict the asymmetric pattern characteristic of non-linear elasticity." They found this earthquake causing three faults: the Lavic Lake fault (LL), the Bullion Fault (BF), and the Bullion Mountain fault (BM, Fig.55b). In the Lavic Lake fault, they found "relatively small internal deformation near the area where the fault changes direction (at N34.6°)". They noticed the loss of phase coherence in "other earthquakes", and attributed this phenomenon "in part to the intense shaking of the ground disrupting surface scatters such as small rocks and gravels". The range displacement profiles along four parallel lines of Fig.55b are shown in Fig.55c. Peltzer et al., (2001) found "a complicated rupture", "internal deformation", "loss of phase coherence", "fault changes direction" etc, but did not explain what caused those phenomenon. The earthquake vapor model could.

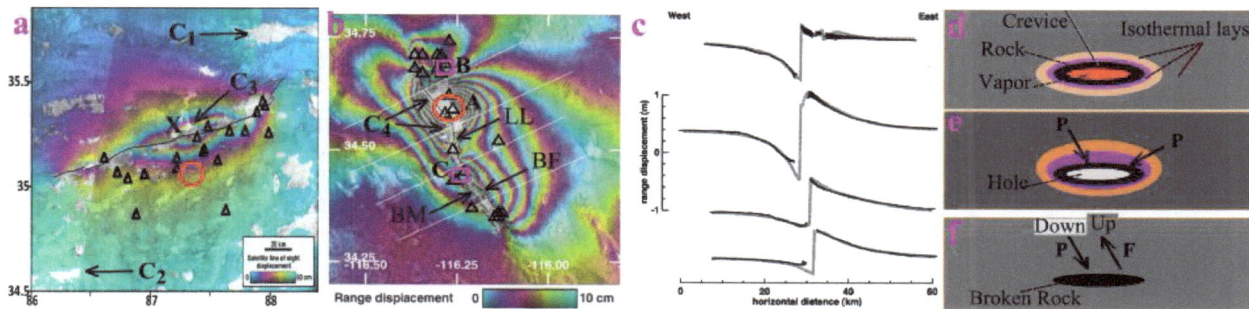

Fig.55 InSAR image and Earthquake vapor explanation

(**a**) A satellite synthetic aperture radar interferometry (InSAR) image originally from Peltzer et al. (1999) reveals surface change around the M7.9 Manyi, Xizang earthquake (35.1, 87.3) on Nov.8, 1997. It consists of three tracks whose data were acquired in 1997 respectively on Mar.16 and Nov.16 (left track), Aug.19 and Dec.2 (middle track), and May 22 and Dec.18 (right track). A full-color "cycle" (blue-red-yellow-blue) represents 50 cm of ground shift away from satellite along the radar line of sight. Black solid curve: extrapolated fault. Red circle and black **x**: the M7.9 epicenter reported by the National Earthquake Information Center (NEIC) and Harvard respectively. Black triangles: all moderate shocks in the image between Mar.16 and Dec.18, 1997 (M4~4.9 on Nov.8~Dec.7 in fact). Earthquake data except the epicenter were reported by the USGS. C_1 and C_2 are clouds, so is C_3 near epicenter X.

(**b**) This InSAR image originally from Peltzer et al. (2001) depicts range displacement from Sep.15 to Oct. 20, 1999 around the M7.4 Hector Mine earthquake (34.6, -116.3) on Oct.16, 1999. A color cycle indicates 10cm ground shift. Black curves: extrapolated Lavic Lake fault (LL), Bullion Fault (BF), and Bullion Mountain fault (BM). Earthquakes M7.4 (red circle A), M5.8 (magenta square B), M5.7 (magenta square C), and all other moderate (M4~4.9) shocks (black triangles) are reported by the USGS. Cloud C_4 near epicenter A is like C_3.

(**c**) Range displacement profiles (Peltzer et al. 2001). Black dots: 1km/pixel. Gray solid line: modeled.

(**d-f**) Schematic diagrams by the Earthquake Vapor Model to account for SAR lobes. In **d**, red center: hot vapor, black ring: large rock, magenta, yellow and grey: isothermal layers, black line: crevice. In **e**, white center: empty hole or cracks after a complete vapor eruption, black ring: dehydrated rock (300~1520°C), Arrow P: pressure from various directions. At the same time, the isothermal layers become hotter, bigger, and denser. In **f**, the rock melts, creeps, breaks, slides, fills the empty hole, and hits the bottom to cause part of the melt rocks beneath to be compacted down and other part to be shifted upward. This causes various phenomena including surface up and down.

Fig.55d~f gives a series of schematic diagrams of thermal distribution and evolution around an impending hypocenter. Fig.55d depicts earthquake vapor (red) with high temperature and high pressure in holes or cracks inside a huge rock (black). The enormous heat conducting through rocks of uniform property (infused with water) makes the rock's vicinity smooth isothermal layers (magenta, yellow, gray). Fig.55e reveals dehydration. After complete vapor eruption, the rock (black) hole almost has no vapor inside (white) to bear outside pressure, and rock's yield strength drops down sharply (Fig.4). Meanwhile, the isothermal layers become hotter, bigger, and denser. Fig.55f shows a collapsed rock (black) or an earthquake. During the dehydration, the rock loses both vapor inside and strength sharply, and thus can no longer bear weight above. Consequently, the weakest part of the rock melts and creeps, which triggers a break in the rock. Materials above then fill the empty hole and impact the bottom. This causes earthquake waves or an earthquake, and an opposite force upward. The above hypothesis can account for many puzzles.

This hypothesis will first attempt to explain what may induce different smooth InSAR lobes. A large earthquake is usually accompanied by a swarm of earthquakes because of similar conditions of these rocks, such as their levels of motion, water penetration, friction, heat, vaporization, and dehydration. Those accompanying earthquakes have their own isothermal layers jointing with those of the large earthquake to form a set of smooth united isothermal layers. These united layers can account for a coincidence between simple InSAR lobes of Fig.55a and a simple distribution of all reported shocks (M≥4) including the Manyi earthquake by the USGS at the same depth of 33km within both area and time of the image. These united layers can also account for another coincidence between complicate InSAR lobes of Fig.55b and a polycentric distribution of all reported shocks (M≥4) including the Hector Mine earthquake by the USGS within both area and time of the image. Major shocks and InSAR lobes distribute around three epicenters: the M7.4 Hector Mine epicenter (A), an M5.8 (B) and an M5.7 (C).

This hypothesis can also account for "relatively small internal deformation near the area where the fault changes direction (at N34.6°)" and "the loss of phase coherence" around an epicenter (Peltzer et al., 2001). An impending hypocenter reaches $300\sim1,520^{\circ}C$ (Harrington and Shou, 2005). Such high temperature can make rock melt, creep, deform, and break, so it is normal for "internal deformation" and "the fault changes direction" at or near the Hector Mine epicenter (34.6, -116.3). After an earthquake, the hypocenter is still hot. Its heat vaporizes underground water. The vapor rises up, and forms fog and cloud near the epicenter, e.g. cloud C_3 (Fig.55a) and C_4 (Fig.55b). Those clouds can interrupt SAR waves or cause "the loss of phase coherence" (Fig.55a~b). The Hector Mine earthquake (Oct 16, 1999) has a depth close to zero, and thus the surface temperature should be very high. This might cause the closest station at Twenty-nine Palms (52km away from the Hector Mine epicenter) to miss surface temperature data from Oct.6 to Oct.18, 1999 (Fig.32d). Our model can explain the big hole around a hypocenter, which can explain the formation of vortexes during the Japan tsunami earthquake (Fig.54i).

The vapor model does not depend on plate, so there is no difficulty in explaining earthquakes internal to a plate. This model does not depend on fault, either, so there is no problem for earthquakes that happen where no big earthquake happened in history, such as the 1975 Haicheng quake, the 1994 Northridge quake, the 2003 Bam quake and so on.

Summary

Some scientists claim that "Earthquakes Cannot Be Predicted" (Geller et al. 1997). or that prediction "in the short term remains, by scientific consensus, is technically impossible" (Hall, 2011). To refute this, we list some copies of Shou's predictions in Appendix. Skeptics are welcome to make random predictions by advancing the time window of his or her predictions to the future, but keeping the location and magnitude windows to be the same to see how often their simulations work. Skeptics are also welcome to explain the phenomena of earthquake vapor we discussed, such as high temperature, high pressure, and vapor eruption, especially the Bam cloud to show how much "consensus" is scientific.

We usually view earthquakes as disasters. Earthquake vapor contains an enormous amount of energy, and causes not only earthquakes (which we have discussed), but also tsunami, flood, drought, snowstorm, hail, mystery air and sea accidents (Shou et al., 2010, Shou 2011). We have not discussed the latter phenomena in this book, but will do so in a different book if time permits. In this book, we have discussed how earthquake vapor and earthquake are generated, and how to predict earthquakes exactly, so we will in theory be able to minimize disastrous consequences of earthquakes. In practice, we need to overcome data problems, especially those associated with satellite data. Only then will we be able to minimize damages from earthquake disasters.

Might we be able to control earthquakes in the distant future? Perhaps. We could use a drill rocket to drill a hole in an impending hypocenter and inject the right amount of water into it. Some of the water would vaporize, thus reducing the temperature of the hypocenter. Some water and vapor would fill the empty space of the hypocenter to bear upside pressure. As a result, the earthquake may be delayed or eliminated.

We might also harness earthquake vapor to generate heat. If the Earth did not have earthquakes, it would freeze. In our opinion, glaciation appeared during period of low earthquake activity. Thus, the existence and disappearance of dinosaurs might relate to the frequent and infrequent activities of earthquakes, respectively. When we inject water to deter earthquakes, we can also pipe the resultant vapor to generate electricity and heat, and recover fresh water.

Recently, heads of states have signed an agreement in Paris to control use of fossil fuels to slow down "global warming". This agreement is good for controlling air pollution. However, it is not clear whether global warming is true, since this conclusion is based on temperature data that have likely been artificially manipulated (e.g. deleting abnormal temperatures). Moreover, does the 2008 China snow support "Global Warming"? Abnormal temperature over a large area can also be caused by vapor eruption.

Table

Table 3: Temperature data losses in Turkey during the vapor eruption in 1999

Lab.	Airport	Country	Lat.	Lon.	Jul.13	Jul.14	Jul.15-27
	Izmit	Turkey	40.7	29.9	M7.8		
a	Izmir	Turkey	38.3	27.1	Loss	Yes	Yes
b	Istanbul	Turkey	41	28.8	Loss	Yes	Yes
c	Ankara	Turkey	40.1	33	Loss	Yes	Yes
d	Trabzon	Turkey	41	39.7	Loss	Yes	Yes
e	Adana	Turkey	37	35.3	Loss	Highest	Yes
f	Dalaman	Turkey	36.7	28.8	Loss	Highest	Yes
g	Antalya	Turkey	36.9	30.7	Loss	Highest	Yes
h	Balikesir	Turkey	39.6	27.9	Loss	Loss	
i	Konya	Turkey	38	32.5	Loss	Loss	No
j	Merzifon	Turkey	40.8	35.6	Loss	Loss	No
k	Gaziantep	Turkey	37.1	37.4	Loss	Loss	No
l	Van	Turkey	38.5	43.3	Loss	Loss	No
1	Eskisehir	Turkey	39.8	30.6	Yes	Loss	No
2	Kayseri	Turkey	38.8	35.4	Yes	Loss	No
3	Malatya	Turkey	38.4	38.1	Yes	Loss	No
4	Diyarbakir	Turkey	37.9	40.2	Yes	Loss	No
5	Erzurum	Turkey	40	41.2	Yes	Loss	No
6	Limnos	Greece	39.9	25.2	Yes	Yes	Yes
7	Alexandroupoli	Greece	40.8	25.9	Yes	Yes	Yes
8	Mytilini	Greece	39.1	26.6	Yes	Yes	Yes
9	Samos	Greece	37.7	26.9	Yes	Yes	Yes
10	Rhodes	Greece	36.4	28.1	Yes	Yes	Yes
11	Bucuresti	Romania	44.5	26.1	Yes	Yes	Yes
12	Varna	Bulgaria	43.2	27.9	Yes	Yes	Yes
13	Odesa	Ukraine	46.4	30.7	Yes	Yes	Yes
14	Larnaca	Cyprus	34.9	33.6	Yes	Yes	Yes
15	Lattakia	Syria	35.5	35.8	Yes	Yes	Yes
16	Cairo	Egypt	30.1	31.4	Yes	Yes	Yes
17	Tel Aviv-Yafo	Israel	32.1	34.8	Yes	Yes	Yes
18	Queen Alia	Jordan	32	36	Yes	Yes	Yes
19	Jeddah	Saudi Arabia	21.7	39.2	Yes	Yes	Yes
20	Kuwait	Kuwait	29.2	48	Yes	Yes	Yes
21	Tbilisi	Georgia	41.7	45	Yes	Yes	Yes
22	Baku	Azerbaijan	40.5	50.1	Yes	Yes	Yes
23	Nicosia	Cyprus	35.2	33.4	No	No	No
24	Simferopol	Ukraine	45	34	No	No	No
25	Sochi	Russia	43.4	39.9	No	No	No
26	Zvartnots	Armenia	40.2	44.4	No	No	No
27	Baghdad	Iraq	33.2	44.2	No	No	No

Note: Lab. Lat. Log. and Jul. are Label, Latitude, Longitude and July respectively. In Column Jul.13, Jul.14 and Jul.15-27, Yes, No, Loss and Highest indicate having data, no data, losing data and the highest data on Jul.14 in 15 years from 1996 to 2010 respectively

Table 4: Abnormal temperatures deleted by the Weather Underground

Date	Time	Temp	Airport	Lat	Lon	Ref
19970602	1:41 pm	300°C	Las Americas	18.4	- 69.7	1
19991117	5:00 am	469°F	Maputo	- 25.9	32.6	2
20031224	11:30 pm	482°F	New Delhi	28.6	77.1	3
20031230	3:30-4:00 pm	1652°F	New Delhi	28.6	77.1	4
20041115	9:30 am	225°C	Karachi	24.9	67.1	5
20041115	9:30 pm	146°C	Lahore	31.5	74.4	6
20041115	10:30-11:30 pm	1500°C	New Delhi	28.6	77.1	7
20041116	12:00-1:30 am	1500°C	New Delhi	28.6	77.1	8
20041116	6:00 pm	288°C	Karachi	24.9	67.1	9
20041117	5:30-11:30 pm	1500°C	New Delhi	28.6	77.1	10
20041117	8:00 pm	246°F	Nairobi Jomo	- 1.3	36.9	11
20041118	12:00 am	242°F	Nairobi Jomo	- 1.3	36.9	12
20041118	12:00-6:00 am	2000°C	New Delhi	28.6	77.1	13
20041118	5:30 pm	205°C	Lahore	31.5	74.4	14
20041215	6:20 pm	141°C	Kerman	30.2	57	15
20051217	7:20 pm	210°F	Kerman	30.2	57	16
20070504	4:02 pm	1342°F	Abidjan	5.2	-3.9	17
20071219	2:20 pm	1830°F	Kerman	30.2	57	18

Note: The Weather Underground website stated, "we assume that any temperature greater than the highest temperature reliably measured on Earth (136F, which is 59°C) is an error, and should be deleted" on Jul.23, 2010. These records in Table 3 were downloaded before Jul.23 and no longer exist. We exhibit them in Google website with the following links.

1 https://docs.google.com/file/d/0B3PS6mjpf0ITRDM1M0gzSFlpQjg/edit?usp=sharing
2 https://docs.google.com/file/d/0B3PS6mjpf0ITS0xQMTk0ZkFSTnc/edit?usp=sharing
3 https://docs.google.com/file/d/0B3PS6mjpf0ITek56MmFkRURDWlk/edit?usp=sharing
4 https://docs.google.com/file/d/0B3PS6mjpf0ITUEh3b0NKUDZPNTA/edit?usp=sharing
5 https://docs.google.com/file/d/0B3PS6mjpf0ITNmhxcUdPNHl2TjQ/edit?usp=sharing
6 https://docs.google.com/file/d/0B3PS6mjpf0ITNEN6eHllLUFNSG8/edit?usp=sharing
7 https://docs.google.com/file/d/0B3PS6mjpf0ITcl9EWVpoZl9yYjg/edit?usp=sharing
8 https://docs.google.com/file/d/0B3PS6mjpf0ITUVZrRVE1bnRDX0k/edit?usp=sharing
9 https://docs.google.com/file/d/0B3PS6mjpf0ITTlhCV3oxNnRncmM/edit?usp=sharing
10 https://docs.google.com/file/d/0B3PS6mjpf0ITTE0wLVpKYjNRZDQ/edit?usp=sharing
11 https://docs.google.com/file/d/0B3PS6mjpf0ITd1pwNlZvZHdKQ00/edit?usp=sharing
12 https://docs.google.com/file/d/0B3PS6mjpf0ITX3NlamlqRENJbDA/edit?usp=sharing
13 https://docs.google.com/file/d/0B3PS6mjpf0ITejRjMnl1dUJrQzA/edit?usp=sharing.
14 https://docs.google.com/file/d/0B3PS6mjpf0ITVVF5Yk9KSW91bGc/edit?usp=sharing
15 https://docs.google.com/file/d/0B3PS6mjpf0ITb2dtZGpWUzdTcm8/edit?usp=sharing
16 https://docs.google.com/file/d/0B3PS6mjpf0ITS1o3S1JmQ0JMLWs/edit?usp=sharing
17 https://docs.google.com/file/d/0B3PS6mjpf0ITLVNlMl9vYkZuY1k/edit?usp=sharing
18 https://docs.google.com/file/d/0B3PS6mjpf0ITUVZrRVE1bnRDX0k/edit?usp=sharing

Table 6 Systematic error for the same Indonesia earthquake on Dec. 26, 2004

Date	Time	Lat.(N)	Lon.(E)	Dep.	Magnitude		Rank	Org.
20041226	1:00:40.0	15	81		mb	7.1	A	LED
20041226	0:59:39.0	14.7	94.8	15	mb	6.9	A	INGV
20041226	0:59:30.2	3.7	85.4	10	mb	6.1	A	SED
20041226	0:59:27.4	11.2	94.1	33	mb	5.9	A	LDG
20041226	0:59:23.6	10.5	94.5	33	mb	6	A	LDG
20041226	0:59:14.2	6.6	93.8	25	mb	5.6	A	NEWS
20041226	0:59:04.9	4.6	93.3	25	mb	5.5	A	NEWS
20041226	0:59:04.0	5.7	95.9	30	mb	7.3	A	ODC
20041226	0:59:00.0	5.9	98		Mw	8.5	A	FLN
20041226	0:59:00.0	5.1	95.5		mb	6.6	A	BRA
20041226	0:58:59.3	2.7	92.6	33	mb	6.3	A	NOR
20041226	0:58:53.0	8.8	98.2		Mw	8	A	ELRO
20041226	0:58:51.0	17	111.2		mb	6.4	A	RNS
20041226	0:58:48.0	10.8	98.8		Mw	8.1	A	ELRO
20041226	0:58:45.0	3.2	97.6		mb	6.7	A	GFZ
20041226	0:58:43.0	3.4	102.7		Mw	8.7	A	EVRO
20041226	0:58:41.0	2.6	97.4		mb	6.9	A	GFZ

Note:

1. All data in this table are from the European-Mediterranean Seismological Centre (http://www.emsc-csem.org/cgi-bin/ALERT_all_messages.sh?1) for the same Sumatra tsunami earthquake on Dec. 26, 2004.

2. Lat. Latitude (N). Lon. Longitude (E). Dep. Depth (km).

3. Org. Organization.

BRA: Seismology Division, Slovak Academy of Sciences in Bratislava, Slovakia

ELRO: Servicio Hydrografico y Oceanografico de la Armada del Chile in Chile

EVRO: Instituto de Ciencas da Terra e do Espaco in Portugal

FLN: Laboratoire de Détection Géophysique in France

GFZ: GeoForschungsZentrum (GEOFON) in Potsdam, Germany

INGV: Italian National Seismic Network in Roma, Italy

LDG: Laboratoire de Détection et de Géophysique in Bruyères-le-Châtel, France

LED: Landsamt für Geologie, Rohstoffe und Bergbau in Baden Wuerttemberg, Germany

NEWS: Norwegian Seismic Array in Kjeller, Norway

NOR: Norwegian Seismic Array in Kjeller, Norway

ODC: Observatories and Research Facilities for EUropean Seismology in De Bilt, The Netherlands

RNS: Réseau National de Surveillance Sismique in Strasbourg, France

SED: Swiss Seismological Service in Zuerich, Switzerland

Table 7 A comparison between two databases demonstrates the problem of missing earthquake data

No.	Date	SI-NOA				PDE				Δlat	Δlon	ΔMs
		Lat	Lon	Ms	Loss	Lat	Lon	Ms	Loss			
1	19870107	40.367	20.800	5.4		40.445	20.584	5.3		-0.1	0.2	0.1
2	19870213	40.100	20.183	5.4		40.200	19.823	5.4		-0.1	0.4	0.0
3	19870227	38.367	20.417	5.9		38.473	20.200	5.6		-0.1	0.2	0.3
4	19870308	39.517	20.350	5.0		39.472	20.569	5.2		0.0	-0.2	-0.2
5	19870412	35.517	23.517	5.4		35.502	23.370	5.4		0.0	0.1	0.0
6	19870514				1	38.227	22.042	5.0				
7	19870529	37.533	21.600	5.5		37.549	21.574	5.5		0.0	0.0	0.0
8	19870610	37.167	21.467	5.5		37.230	21.465	5.5		-0.1	0.0	0.0
9	19870806	39.217	26.250	5.2					1			
10	19870818	40.217	25.017	5.3					1			
11	19870827	38.933	23.817	5.3		38.904	23.757	5.1		0.0	0.1	0.2
12	19870915				1	37.852	26.934	5.0				
13	19871210	35.400	26.633	5.2					1			
14	19871210	36.650	21.683	5.2		36.634	21.680	5.0		0.0	0.0	0.2
15	19871213				1	37.217	20.475	5.0				
16	19880109	41.217	19.667	5.6		41.246	19.630	5.6		0.0	0.0	0.0
17	19880109	35.817	21.733	5.3		35.826	21.739	5.1		0.0	0.0	0.2
18	19880122	38.633	21.017	5.1		38.642	20.980	5.2		0.0	0.0	-0.1
19	19880218	39.117	23.467	5.1					1			
20	19880308	38.817	21.117	5.1					1			
21	19880326	40.083	19.850	5.4		40.180	19.890	5.3		-0.1	0.0	0.1
22	19880424	38.883	20.333	5.0					1			
23	19880509				1	37.708	19.965	5.0				
24	19880514	41.833	19.633	5.2					1			
25	19880518	38.350	20.467	5.8		38.418	20.479	5.7		-0.1	0.0	0.1
26	19880522	38.350	20.533	5.5		38.409	20.464	5.3		-0.1	0.1	0.2
27	19880602	38.267	20.367	5.0					1			
28	19880606	38.300	20.483	5.0		38.393	20.465	5.0		-0.1	0.0	0.0
29	19880705				1	38.146	22.845	5.3				
30	19880712	38.783	23.433	5.0					1			
31	19880911	38.150	23.217	5.0					1			
32	19880922	37.983	21.117	5.5		38.022	21.089	5.3		0.0	0.0	0.2
33	19881014	40.167	19.517	5.1		40.152	19.740	5.0		0.0	-0.2	0.1
34	19881016	37.900	20.967	6.0		37.983	20.932	5.8		-0.1	0.0	0.2
35	19881108				1	36.575	22.659	5.2				
36	19881213				1	37.846	21.193	5.0				
37	19881214	39.767	20.317	5.1					1			
38	19881222	38.333	21.750	5.0					1			
39	19890126	40.317	19.033	5.0					1			
40	19890226				1	37.196	20.800	5.1				
41	19890226				1	39.155	24.506	5.0				
42	19890314				1	35.508	23.322	5.0				
43	19890317	41.383	19.700	5.4		41.237	19.891	5.4		0.1	-0.2	0.0
44	19890319	39.283	23.583	5.8		39.254	23.516	5.5		0.0	0.1	0.3

No	Date	Lat	Lon	M		Lat	Lon	M				
45	19890319	39.233	23.633	5.0					1			
46	19890428	39.267	23.567	5.2					1			
47	19890501	37.183	21.233	5.1		37.212	21.145	5.1		0.0	0.1	0.0
48	19890515				1	38.311	21.816	5.1				
49	19890607	38.000	21.633	5.2		38.057	21.620	5.3		-0.1	0.0	-0.1
50	19890714	41.750	20.217	5.3		41.935	20.020	5.0		-0.2	0.2	0.3
51	19890801	39.200	23.633	5.0					1			
52	19890815	39.167	26.217	5.3					1			
53	19890820	37.217	21.083	5.9		37.278	21.280	5.7		-0.1	-0.2	0.2
54	19890824	37.917	20.117	5.7		37.995	20.183	5.4		-0.1	-0.1	0.3
55	19890824				1	37.949	20.104	5.0				
56	19890905	40.150	25.067	5.4		40.200	25.086	5.2		-0.1	0.0	0.2
57	19890919	39.483	21.333	5.0					1			
58	19891124	36.733	26.633	5.3					1			
	sum or max				12				19	0.2	0.4	0.3

Note: Geller (1996), editor of Geophysical Research Letters (GRL), represented GRL to offer the public two catalogs to comment VAN's method. One catalog was from the Greek Seismological Institute-National Observatory of Athens (SI-NOA), and the other was from NOAA (PDE) for earthquakes of M≥5.0 in (35~42, 17~27) and 1987~1989. In together 58 data, the SI-NOA lost 12 (20.7%), and the NOAA lost 19 (32.8%). In 27 common data, the maximum errors of latitudes, longitudes and magnitudes are 0.2, 0.4 and 0.3 respectively.

Table 8: Earthquakes of death toll ≥ 10,000 in 1990~2015

Date	Time	Lat.	Lon.	M.	Location	Death	Vapor
19900620	21:00	36.96	49.41	7.7	Rasht, Iran	50,000	Fig.30a
19990817	1:39	40.75	29.86	7.7	Izmit, Turkey	17,118	Fig.17
20010126	3:16	23.42	70.23	8.0	Gujarat, India	20,085	Fig.37
20031226	1:56	28.99	58.29	6.8	Bam, Iran	31,000	Fig.6, 7
20041226	0:58	3.32	95.85	9.0	Sumatra, Indonesia	227,898	Fig.13
20051008	3:50	34.43	73.54	7.6	Kashmir, Pakistan	86,000	Fig.38
20080512	6:28	31.10	103.28	8.0	Sichuan, China	87,587	Fig.39
20100112	21:53	18.44	-72.54	7.3	Haiti	316,000	Fig.40
20110311	5:46	38.30	142.37	9.1	Honshu, Japan	28,050	Fig.41

Note: Lat., Lon, and M: latitude, longitude and magnitude respectively. The data were from the USGS.

Table 9: Increase in daily maximum temperatures during the Honshu eruption on Feb.23~25

Name	Lat.	Lon.	Feb.22	Feb.23	Feb.24	Feb.25	Max	Δ
Kagoshima	31.8	130.7	17.0	15.0	18.0	20.0	20.0	3.0
Kitakyushu	33.8	131	10.0	11.0	12.0	16.0	16.0	6.0
Busan	35.1	129	12.2	12.8	16.7	16.1	16.7	4.4
Tokyo	35.5	139.8	10.0	11.0	14.0	20.0	20.0	10.0
Sendai	38.1	140.9	7.8	8.3	12.0	13.9	13.9	6.1
Akita	39.6	140.2	8.0	10.0	12.0	6.0	12.0	4.0
Aomori	40.7	140.7	5.0	4.0	11.0	3.0	11.0	6.0
Hakodate	41.8	140.8	5.0	7.0	10.0	4.0	10.0	5.0
Obihiro	42.7	143.2	2.0	0.0	5.0	4.0	5.0	3.0
Chitose	42.8	141.7	4.0	3.0	6.0	3.0	6.0	2.0
Yuzhno	46.9	142.7	-1.0	1.0	3.0	1.0	3.0	4.0
Khabarovsk	48.5	135.2	-2.0	3.0	-2.0	-12.0	3.0	5.0
Fukuoka	33.6	130.4	13.0	16.0	20.0	15.6	20.0	7.0
Osaka	34.7	135.5	16.0	16.0	17.0	18.0	18.0	2.0
Vladivostok	43.1	131.9	3.0	5.0	4.0	-5.0	5.0	2.0
Kushiro	43	144.2	2.0	1.0	2.0	7.0	7.0	5.0
Magadan	59.6	150.8	-21.1	-19.4	-18.9	-12.0	-12.0	9.1

Note: Lat.: Latitude, Lon. : Longitude, Max.: Maximum daily maximum temperature during the Honshu eruption on Feb.23~25. Δ: Difference between Column Feb.22 and Max. The average temperature increase is 4.9°C. The 17 airports cover an area of 2.06 million square kilometer.

Table10 All earthquake predictions submitted to the USGS and their statistical evaluations

		Earthquake		Predictions		Associated		Earthquakes			Acc	Prior Probabilities				Adj.	
No.	D	Signed	End	Location	M	Date	Time	Lat	Lon	M	H	P_C	Af	P_{RJ}	P_{Comb}	Score	Var
1		940214	940310	Around Pas (33~35, -119~-117)	4~5.5	940225	1259	34.4	-118.5	4.1	1	0.147	0		0.15	1.77	0.54
2		940308	940330	Mexico, *S Cal	5.5~6.8	940312	2346	16.7	-94.3	5.6	1	0.449	0		0.45	0.77	0.48
3		940316	940409	Around Pas (33~35, -119~-117)	4~5.5	940320	2120	34.2	-118.5	5.3	1	0.147	1	0.59	0.65	0.52	0.50
D1	D2	940316	940409	*Jap, Kuril or NW China, Pamir	5.5~6.8	940406	519	36.2	141.5	5.5	1	0.916	0		0.92	0.22	0.51
D2	D1	940321	940409	30~45,65~80	5~6.8	940501	1200	36.9	67.2	6.3	0	0.731	0		0.73	-1.19	0.52
4		940331	940424	Cal	5~7	940406	1901	34.2	-117.1	5.0	1	0.254	0		0.25	1.24	0.52
5		940426	940518	*N Mex, S Cal (23.6~37N)	≥4	940512	22	25.0	-109.3	5.6	1	0.897	0		0.90	0.25	0.52
6		940603	940628	S Cal	3.7~5.5	940615	559	34.3	-118.4	4.2	1	0.833	1	0.52	0.92	0.21	0.50
7		940914	940925	20-50, 0-75	≥6	941025	54	36.4	71.0	6.2	0	0.221	1	0.00	0.22	-0.39	0.53
8		940916	941011	Jap-Aleutian<500km	≥5	940918	1649	38.7	142.9	5.0	1	1.000	0		1.00	0.00	0.00
9		941018	941112	USA	≥5	941027	1745	43.5	-127.4	6.3	1	0.495	0		0.50	0.70	0.48
10		950308	950401	Mexico, S Cal	≥4	950310	706	15.0	-92.8	4.5	1	0.999	0		1.00	0.01	0.05
11		950630	950720	S Cal	≥5	950630	1158	24.7	-110.2	6.2	0	0.120	0		0.12	-0.27	0.53
12		951011	951105	Cal	≥5	951021	238	16.8	-93.5	7.2	0	0.294	0		0.29	-0.46	0.51
13		960510	960530	S Cal (≤37N)	3.7~5.3	960521	2050	37.4	-121.7	4.8	0	0.768	0		0.77	-1.32	0.53
14		961025	961119	S Cal	≥4.5	961127	2017	36.1	-117.7	5.3	0	0.303	0		0.30	-0.47	0.51
15		961125	961220	Mex~Peru	≥6	961231	1241	15.8	-93.0	6.4	0	0.517	0		0.52	-0.72	0.48
D3	15	961209	970105	Mex	≥4.5	961210	2031	16.1	-96.7	4.8	1	0.987	0		0.99	0.06	0.24
16		961205	961229	S Cal,N Mex >30N	4~5.3	961217	403	36.1	-117.7	4.0	1	0.778	1	0.16	0.81	0.35	0.54
17		970306	970405	N China (>35.8N)	≥6	970405	2346	39.5	76.9	5.9							
			LT 20:36			970406	436	39.5	77.0	6.0	1	0.042	0		0.04	3.08	0.42
18		970424	970610	S Cal	≥4	970426	1037	34.4	-118.7	5.1	1	0.861	1	0.01	0.86	0.29	0.54
19		970428	970611	S Cal	3.7~5.3	970506	1912	35.5	-118.4	4.5	1	0.958	0		0.96	0.13	0.42
20		970509	970608	S Cal	4~5.3	970524	436	35.8	-117.6	4.0	1	0.712	1	0.02	0.72	0.45	0.52
21		970528	970712	Turkey-Med sea ≥15E	≥5.5	970727	1007	35.6	21.1	5.8	0	0.266	0		0.27	-0.43	0.52
22		970718	970809	S Cal	≥4	970726	314	33.4	-116.4	4.8	1	0.581	0		0.58	0.59	0.49
23		970804	970829	S Cal	≥4	970806	1104	37.0	-121.5	4.0	1	0.641	0		0.64	0.53	0.50
24		980105	980218	25~41,53~105	≥6	980204	1433	37.1	70.1	6.1	1	0.450	0		0.45	0.77	0.48
25		980107	980220	Mex	≥5	980203	302	15.9	-96.3	6.4	1	0.909	0		0.91	0.23	0.51
26		980309	980423	15~30,<-150	≥4	980507	2315	19.2	-155.5	4.3	0	0.495	0		0.50	-0.69	0.48
27		980406	980522	Mex, Cal <34N	≥4.5	980408	402	16.0	-95.7	5.0	1	1.000	0		1.00	0.00	0.00
28		980724	980902	34~39,-119~-117	4~5.5	980801	601	37.6	-118.8	4.4	1	0.595	1	0.12	0.64	0.52	0.50
29		981123	990109	Cal <39N	≥4.5	981212	141	37.5	-116.3	4.5	1	0.586	0		0.59	0.59	0.49
30		981228	990213	33~39, -120~-116	4.2~5.4	990127	1044	36.8	-116.0	4.8	1	0.735	0		0.74	0.43	0.52
31		980222	990408	20~38, 50~100	≥5.5	990304	538	28.3	57.2	6.6	1	0.818	0		0.82	0.35	0.54
32		990402	990520	24~34, -118~-108	4~5.2	990407	626	32.6	-116.2	4.0	1	0.919	1	0.02	0.92	0.21	0.50
33		990412	990529	34~39,≤-116	≥4	990514	754	34.1	-116.4	4.9	1	0.945	1	0.18	0.95	0.14	0.43
						990514	1052	34.0	-116.4	4.2							
34		990505	990621	27~33, -117~-113	≥4	990601	1518	32.4	-115.2	5.1	1	0.656	0		0.66	0.51	0.50
D4	34	990517	990704	Mex <29N	≥5	990615	2042	18.4	-97.4	7.0	1	0.911	0		0.91	0.22	0.51
35		990609	990725	35~39, -120~-116	4~5.3	990711	1820	35.7	-118.5	4.6	1	0.634	0		0.63	0.53	0.50
36		990726	990910	36~42, 113~117	≥5	991101	1325	39.9	114.0	5.5	0	0.029	0		0.03	-0.10	0.36
37		990825	991003	N Cal >38, <-122	≥5.5	990922	2227	38.4	-122.6	4.3	0	0.088	0		0.09	-0.22	0.51
						990818	106	37.9	-122.7	5.0							
38		990927	991114	Turkm or Caspian <41, <56	≥5.5	991112	1657	40.8	31.2	7.5	0	0.037	0		0.04	-0.12	0.40
D5	38	991025	991114	30~40,51~58	≥6	991112	1657	40.8	31.2	7.5	0	0.015	0		0.02	-0.06	0.26
39		991004	991118	34~41, -5~1	≥5	991222	1736	35.3	-1.3	5.7	0	0.087	0		0.09	-0.22	0.51
D6	39	991028		Cancel							1						
40		991028	991214	30~33, -117~-115	≥4.3	991121	646	18.5	-107.2	6.2	0	0.317	0		0.32	-0.49	0.51
41		991227	000211	Indian Ocean >20S	≥7	000209	1840	-27.6	65.7	5.0	0	0.000	0		0.00	0.00	0.00
						000209	1840	-27.7	65.7	5.0							
						000210	1418	-27.6	65.7	5.7							
						000210	1418	-27.7	65.7	5.7							
						000210	2300	-27.6	65.8	5.5							
						000210	2300	-27.6	65.8	5.6							
42		000131	000310	S Iran <30, 58~68	≥4.5	000217	944	29.6	67.1	4.6	1	0.170	0		0.17	1.63	0.54
43		000224	000310	S Iran <32, 57~58	≥4.5	000229	1716	28.2	57.1	4.5	1	0.212	1	0.03	0.24	1.31	0.53
44		000228	000413	31~35, -116.5~-115	≥4.5	000409	1048	32.7	-115.4	4.3	0	0.292	0		0.29	-0.46	0.51
45		000228	000418	36.5~38.5, 36~39	1M5/2M4	000402	1141	37.6	37.3	4.5	1	0.095	0		0.10	2.22	0.52
						000402	1726	37.6	37.4	4.3							
46		000322	000505	35.8~40.7, -120~-117	≥4	000328	1516	36.0	-117.9	4.3	1	0.564	0		0.56	0.61	0.48
47		000418	000606	62~65, -27~-20	≥3.5	000617	1540	64.0	-20.5	6.8	0	0.263	0		0.26	-0.43	0.52
48		000419	000604	33~35, -119.5~-115.5	≥4	000510	2325	33.2	-115.6	3.9	0	0.602	1	0.00	0.60	-0.86	0.49
D7	48	000421	000604	33.5~36.5, -118~-115	≥4	000527	335	35.8	-117.7	4.0	1	0.539	1	0.01	0.54	0.64	0.48
D8	48	000421	000604	34~35.5, -119.5~-118	≥4	000523	442	36.3	-118.1	4.0	0	0.185	0		0.19	-0.35	0.54
D9	48	000425	000605	33.5~37, -118~-115	≥4	000527	335	35.8	-117.7	4.0	1	0.561	1	0.01	0.57	0.61	0.48
49		000428	000615	-35~-25, 60~85	1M5/2M4	000617	2259	-28.5	62.8	4.5	0	0.726	0		0.73	-1.17	0.52
						000624	2221	-28.5	62.7	4.2							
D10	49	000531		Cancel							1						
50		000619	000807	Cal Neighbor<39	≥5.5	000721	613	18.4	-98.9	5.9	1	0.726	0		0.73	0.44	0.52
D11	50	000626	000807	Cal Neighbor	≥5.5	000721	613	18.4	-98.9	5.9	1	0.744	0		0.74	0.42	0.52
D12	50	000714	000807	32~39, <-114	≥5	000903	836	38.4	-122.4	5.2	0	0.243	0		0.24	-0.41	0.53
51		000629	000820	Jap <37	≥6	000701	701	34.2	139.1	6.2	1	0.486	0		0.49	0.71	0.48
52		000705	000821	Jap, E China sea<34, <142.5	≥6	000730	1225	33.9	139.4	6.5	1	0.378	0		0.38	0.90	0.49
53		001220	010304	35~42,-106~-104	≥4	010223	2143	38.7	-112.6	4.1	0	0.099	0		0.10	-0.24	0.52
D13	53	010302	010402	32~42,-108~-103	≥4	010223	2143	38.7	-112.6	4.1	0	0.110	0		0.11	-0.26	0.53
D14	D13	010329		Cancel							1						
54		010308	010525	36~43, 27~34	≥5.5	010524	318	39.3	27.9	4.5	0	0.157	0		0.16	-0.32	0.54
						010524	625	39.4	27.8	4.4							
D15	54	010314	010523	36~43, 26~32	≥6	010524	318	39.3	27.9	4.5	0	0.061	0		0.06	-0.17	0.47
						010524	625	39.4	27.8	4.4							
D16	D15	010320		Cancel							1						
55		010309	010522	10~36, 90~107	≥6.5	010315	39	8.7	94.0	5.4	0	0.142	0		0.14	-0.30	0.54
						010315	122	8.7	94.0	6.0							
D17	55	010316		Cancel							1						
56		010320	010504	N Cal (37~42), -126~-122	≥4.5	010420	519	40.7	-125.3	4.8	1	0.365	0		0.37	0.93	0.50
D18	56	010321	010505	Cal >38-	≥4	010322	2122	40.5	-126.2	4.7	1	0.844	0		0.84	0.32	0.54
D19	56	010329	010518	Cal, Nev	≥5	010717	1207	36.0	-117.9	5.2	0	0.538	0		0.54	-0.75	0.48
D20	56	010330	010519	33~38, <-116	≥5	010717	1207	36.0	-117.9	5.2	0	0.326	0		0.33	-0.49	0.50
D21	56	010402	010522	33~38, <-116	1M5/2M4	010517	2153	35.8	-118.0	4.0	1	0.635	0		0.63	0.53	0.50
						010517	2256	35.8	-118.0	4.2							
57		010403	010702	36.3~37.2, -121.5~-120	≥4	010702	1733	36.7	-121.3	4.1	1	0.310	0		0.31	1.06	0.51
58		010405	010522	36~43, 25~36	≥5.5	010610	1311	38.6	25.6	5.6	0	0.151	0		0.15	-0.31	0.54
59		010424	010601	Cal, Nev <38, >-121	1M5/2M4	010517	2153	35.8	-118.0	4.0	1	0.595	0		0.60	0.58	0.49
						010517	2256	35.8	-118.0	4.2							
D22	59	010509	010609	Cal,Nev,Mex>26,<-112	≥5	010810	2019	39.8	-120.6	5.2	0	0.457	0		0.46	-0.64	0.48
60		010426	010615	USA,Can 38~54, <-120	1M5/2M4	010502	205	49.9	-130.2	5.3	1	0.740	0		0.74	0.43	0.52
61		010716	010929	48.4~53,-120~-112	≥6	010914	445	48.7	-128.7	6.0	0	0.000	0		0.00	0.00	0.00
D23	61	010801	010929	Can, USA (48.4~53,<-112)	≥6	010914	445	48.7	-128.7	6.0	1	0.158	0		0.16	1.70	0.54
62		010805	011021	21~25, 68~73	≥6	010902	225	0.9	82.5	6.1	0	0.010	0		0.01	-0.05	0.21
63		010810	011002	32~33.5, -117~-115.2	≥4	011031	756	33.5	-116.5	5.2	0	0.489	0		0.49	-0.68	0.48
Sum											52		13			15.7	37.5

Overall P-value 0.005 Overall accuracy (%) 60.5 Z score 2.56

Note: D1~23: Dependent predictions (see main text for definition). Other 63 predictions are independent. Singed: by the USGS. M: Magnitude. Lat: Latitude. Lon: Longitude. Acc: Accuracy. H: hit (1: yes; 0: no). Af: Aftershock (1: yes; 0: no). *Pc*: Catalog probability. P_{RJ}: Aftershock probability. *Pcomb*: Combined probability. Adj. Score: score adjusted so that no skill has a score of zero. Var: Variance.

In column "Location", Cal: California, Can: Canada, Jap: Japan, Med: Mediterranean, Mex: Mexico, Nev: Nevada, Pas: Pasadena and Turkm: Turkmenistan, ,

Suppose earthquake data to be complete and without errors and an earthquake to be at a point, all predictions (including dependent) have an accuracy of 60%m and a P-value 0.005, which proves this set of predictions significantly outperforming "no skill".

Table11. Evaluation of 63 independent predictions

No.	\<Predictions\> Begin	End	Location	M	\<Associated Earthquake\> Date	Time	Lat	Lon	M	\<Accuracy\> T	L	M	all	\<Prior Probability\> Pc	Af	PRJ	PComb	\<Adj.\> Score	Var
1	940213	940310	Around Pas (33~35, -119~-117)	4~5.5	940225	1259	34.4	-118.5	4.1				1	0.147	0		0.147	1.771	0.541
2	940305	940330	Mex, S Cal	5.5~6.8	940312	2346	16.7	-94.3	5.6				1	0.449	0		0.449	0.770	0.483
3	940315	940409	Around Pas (33~35, -119~-117)	4~5.5	940320	2120	34.2	-118.5	5.3				1	0.147	1	0.591	0.651	0.517	0.499
4	940330	940424	California	5~7	940406	1901	34.2	-117.1	5.0				1	0.254	0		0.254	1.241	0.524
5	940423	940518	N Mex, S Cal (23.6~37N)	≥4	940512	22	25.0	-109.3	5.6				1	0.897	0		0.897	0.245	0.524
6	940603	940628	S Cal	3.7~5.5	940615	559	34.3	-118.4	4.2				1	0.833	1	0.523	0.920	0.208	0.500
7	940910	940925	20-50, 0-75	≥6	941025	54	36.4	71.0	6.2	0	1	1	0	0.221	1	0.002	0.222	-0.390	0.533
8	940916	941011	Jap-Aleutian<500km	≥5	940918	1649	38.7	142.9	5.0				1	1.000	0		1.000	0.000	0.000
9	941018	941112	USA	≥5	941027	1745	43.5	-127.4	6.3				1	0.495	0		0.495	0.700	0.480
10	950307	950401	Mex, S Cal	≥4	950310	706	15.0	-92.8	4.5				1	0.999	0		0.999	0.007	0.048
11	950630	950720	S Cal	≥5	950630	1158	24.7	-110.2	6.2	1	0	1	0	0.120	0		0.120	-0.270	0.534
12	951011	951105	Cal	≥5	951021	238	16.8	-93.5	7.2	1	0	1	0	0.294	0		0.294	-0.462	0.513
13	960510	960530	S Cal (≤37N)	3.7~5.3	960521	2050	37.4	-121.7	4.8	1	*0	1	0	0.768	0		0.768	-1.325	0.530
14	961025	961119	S Cal	≥4.5	961127	2017	36.1	-117.7	5.3	0	1	1	0	0.303	0		0.303	-0.471	0.511
15	961125	961220	Mex~Peru	≥6	961231	1241	15.8	-93.0	6.4	0	1	1	0	0.517	0		0.517	-0.717	0.481
16	961204	961229	S Cal, N Mex >30N	4~5.3	961217	403	36.1	-117.7	4.0				1	0.778	1	0.162	0.814	0.351	0.540
17	970306	970405	N China (>35.8N) LT 20:36	≥6	970405	2346	39.5	76.9	5.9										
					970406	436	39.5	77.0	6.0				1	0.042	0		0.042	3.078	0.415
18	970424	970610	S Cal	≥4	970426	1037	34.4	-118.7	5.1				1	0.861	1	0.012	0.863	0.293	0.539
19	970427	970611	S Cal	3.7~5.3	970506	1912	35.5	-118.4	4.5				1	0.958	0		0.958	0.135	0.415
20	970508	970608	S Cal	4~5.3	970524	436	35.8	-117.6	4.0				1	0.712	1	0.023	0.719	0.450	0.517
21	970528	970712	Turkey-Med sea ≥15E	≥5.5	970727	1007	35.6	21.1	5.8	0	1	1	0	0.266	0		0.266	-0.435	0.521
22	970719	970809	S Cal	≥4	970726	314	33.4	-116.4	4.8				1	0.581	0		0.581	0.592	0.486
23	970804	970829	S Cal	≥4	970806	1104	37.0	-121.5	4.0				1	0.641	0		0.641	0.527	0.497
24	980105	980218	25~41, 53~105	≥6	980204	1433	37.1	70.1	6.1				1	0.450	0		0.450	0.768	0.483
25	980106	980220	Mex	≥5	980203	302	15.9	-96.3	6.4				1	0.909	0		0.909	0.227	0.514
26	980309	980423	15~30,<-150	≥4	980507	2315	19.2	-155.5	4.3	0	1	1	0	0.495	0		0.495	-0.686	0.480
27	980406	980522	Mex, Cal <34N	≥4.5	980408	402	16.0	-95.7	5.0				1	1.000	0		1.000	0.000	0.000
28	980724	980902	34~39, -119~-117	4~5.5	980801	601	37.6	-118.8	4.4				1	0.595	1	0.122	0.644	0.524	0.497
29	981123	990109	Cal <39N	≥4.5	981212	141	37.5	-116.3	4.5				1	0.586	0		0.586	0.586	0.487
30	981228	990213	33~39, -120~-116	4.2~5.4	990101	1044	36.8	-116.0	4.8				1	0.735	0		0.735	0.434	0.521
31	990222	990408	20~38, 50~100	≥5.5	990304	538	28.3	57.2	6.6				1	0.818	0		0.818	0.347	0.540
32	990402	990520	24~34, -118~-108	4~5.2	990407	626	32.6	-116.2	4.0				1	0.919	1	0.016	0.920	0.208	0.501
33	990412	990525	≤-116	≥4	990514	754	34.1	-116.4	4.9				1	0.945	1	0.176	0.955	0.142	0.427
34	990505	990621	27~33, -117~-113	≥4	990601	1518	32.4	-115.2	5.1				1	0.656	0		0.656	0.512	0.500
35	990609	990725	35~39, -120~-116	4~5.3	990711	1820	35.7	-118.5	4.6				1	0.634	0		0.634	0.535	0.495
36	990726	990910	36~42, 113~117	≥5	991101	1325	39.9	110.4	5.5	0	1	1	0	0.029	0		0.029	-0.104	0.359
37	990825	991003	N Cal >38, <-122	≥5.5	990922	2227	38.4	-122.6	4.3	1	1	#0	0	0.088	0		0.088	-0.222	0.511
					990818	106	37.9	-122.7	5										
38	990927	991114	Turkm or Caspian <41, <56	≥5.5	991112	1657	40.8	31.2	7.5	1	0	1	0	0.037	0		0.037	-0.123	0.396
39	991003	991118	34~41, -5~1	≥5	991222	1736	35.3	-1.3	5.7	0	1	1	0	0.087	0		0.087	-0.220	0.510
40	991028	991214	30~33, -117~-115	≥4.3	991121	646	18.5	-107.2	6.2	1	0	1	0	0.317	0		0.317	-0.485	0.507
41	991227	000211	Indian Ocean >20S	≥7	000209	1840	-27.6	65.7	5.0	1	1	#0	0	0.000	0		0.000	0.000	0.000
					000209	1840	-27.7	65.7	5.0										
					000210	1418	-27.6	65.7	5.7										
					000210	1418	-27.7	65.7	5.7										
					000210	2300	-27.6	65.8	5.5										
					000210	2300	-27.6	65.8	5.6										
42	000131	000310	S Iran <30, 58~68	≥4.5	000217	944	29.6	67.1	4.6				1	0.170	0		0.170	1.625	0.541
43	000218	000310	S Iran <32, 57~58	≥4.5	000229	1716	28.2	57.1	4.5				1	0.212	1	0.030	0.235	1.311	0.529
44	000228	000413	31~35, -116.5~-115	≥4.5	000409	1048	32.7	-115.4	4.3	1	1	*0	0	0.292	0		0.292	-0.460	0.514
45	000228	000418	36.5~38.5, 36~39	1M5/2M4	000402	1141	37.6	37.3	4.5				1	0.095	0		0.095	2.221	0.518
					000402	1726	37.6	37.4	4.3										
46	000322	000505	35.8~40.7, -120~-117	≥4	000328	1516	36.0	-117.9	4.3				1	0.564	0		0.564	0.612	0.484
47	000418	000606	62~65, -27~-20	≥3.5	000617	1540	64.0	-20.5	6.8	0	1	1	0	0.263	0		0.263	-0.432	0.522
48	000419	000604	33~35, -119.5~-115.5	≥4	000510	2325	33.2	-115.6	3.9	1	1	*0	0	0.602	1	0.000	0.602	-0.860	0.489
49	000428	000615	25~35S, 60~85E	1M5/2M4	000617	2259	-28.5	62.8	4.5	0	1	1	0	0.726	0		0.726	-1.172	0.519
					000624	2221	-28.5	62.7	4.2										
50	000619	000807	Cal, Neighbor<39N	≥5.5	000721	613	18.4	-98.9	5.9				1	0.726	0		0.726	0.442	0.519
51	000629	000820	Jap <37N	≥6	000701	701	34.2	139.1	6.2				1	0.486	0		0.486	0.713	0.481
52	000705	000821	Jap E China sea<34, <142.5	≥6	000730	1225	33.9	139.4	6.5				1	0.378	0		0.378	0.900	0.493
53	001217	010304	35~42, -106~-104	≥4	010223	2143	38.7	-112.6	4.1	1	0	1	0	0.099	0		0.099	-0.239	0.521
54	010308	010525	36~43, 27~34	≥5.5	010524	318	39.3	27.9	4.5	1	1	#0	0	0.157	0		0.157	-0.318	0.541
					010524	625	39.4	27.8	4.4										
55	010309	010522	10~36, 90~107	≥6.5	010315	39	8.7	94.0	5.4	1	0	#0	0	0.142	0		0.142	-0.299	0.540
					010315	122	8.7	94.0	6.0										
56	010320	010504	N Cal (37~42), -126~-122	≥4.5	010420	519	40.7	-125.3	4.8				1	0.365	0		0.365	0.928	0.495
57	010403	010702	36.3~37.2, -121.5~-120	≥4	010702	1733	36.7	-121.3	4.1				1	0.310	0		0.310	1.064	0.509
58	010405	010522	36~43, 25~36	≥5.5	010611	1311	38.6	25.6	5.6	0	1	1	0	0.151	0		0.151	-0.310	0.541
59	010422	010601	Cal, Nev <38, >-121	1M5/2M4	010517	2153	35.8	-118.0	4.0				1	0.595	0		0.595	0.576	0.488
					010517	2256	35.8	-118.0	4.2										
60	010426	010615	USA,Can 38~54, <-120	1M5/2M4	010502	205	49.9	-130.2	5.3				1	0.740	0		0.740	0.429	0.523
61	010716	010929	Can, USA (48.4~53,-120~-112)	≥6	010914	445	48.7	-128.7	6.0	1	0	1	0	0.000	0		0.000	0.000	0.000
62	010806	011021	21~25, 68~73	≥6	010902	225	0.9	82.5	6.1	1	0	1	0	0.010	0		0.010	-0.046	0.211
63	010808	011002	32~33.5, -117~-115.2	≥4	011031	756	33.5	-116.5	5.2	0	1	1	0	0.489	0		0.489	-0.678	0.481

Notes:
Sum ..... 38 ..... 11 ..... 15.3 28.8

Accuracy (%) 60.3   Z-score 2.84

P-Value 0.002

★ Predictions missed by a small amount.

In these predictions, instead of a large predicted earthquake, a swarm of smaller quakes occurred.

Note: Lat: Latitude. Lon: Longitude: M. Magnitude. T: Time. L: Location. All: product of time T, location L and magnitude M (1: hit. 0: miss). *Pc*: Prediction probability according to the catalog of the USGS. Af: Aftershock (1:

yes. 0: no). P_{R-J}: Aftershock probability according to Reasenberg and Jones. P_{comb}: Combining Pc and P_{R-J} according to Jones and Jones. Adj-Score: Adjusted score according to Jones and Jones. Var: Variance. In column "Location", abbreviations e.g. Cal, Can and so on equal to them in Table 10. Suppose earthquake data without error and miss, and earthquake at a point, the accuracy is **60%** and **P**-Value is **0.002**.

Table 12 Aftershock probability

No.	Associated Earthquake					Acc	Potential Main Shock					Af prob.
	Date	Time	Lat	Lon	M		Date	Time	Lat	Lon	M	
3	19940320	2120	34.2	-118.5	5.3	1	19940117	1230	34.21	-118.53	6.8	0.591
6	19940615	559	34.3	-118.4	4.2	1	19940117	1230	34.21	-118.53	6.8	0.523
							19940525	1256	34.31	-118.39	4.5	0.055
7	19941025	54	36.4	71.0	6.2	0	19930809	1242	36.37	70.86	7.0	0.002
							19910714	909	36.33	71.11	6.7	0.000
16	19961217	403	36.1	-117.7	4.0	1	19961127	2017	36.07	-117.65	5.3	0.162
18	19970426	1037	34.4	-118.7	5.1	1	19940117	2333	34.32	-118.69	5.9	0.012
							19940118	43	34.37	-118.69	5.5	0.005
							19940119	2109	34.37	-118.71	5.5	0.005
							19950626	840	34.39	-118.66	5.2	0.005
20	19970524	436	35.8	-117.6	4.0	1	19950920	2327	35.76	-117.63	6.1	0.023
							19950925	447	35.80	-117.61	5.3	0.004
							19960107	1432	35.76	-117.64	5.4	0.007
							19960108	1052	35.78	-117.63	5.0	0.003
28	19980801	601	37.6	-118.8	4.4	1	19980715	453	37.56	-118.80	5.1	0.122
							19980609	524	37.58	-118.79	5.2	0.054
32	19990407	626	32.6	-116.2	4.0	1	19990313	1331	32.58	-116.16	4.3	0.016
							19990219	308	32.59	-116.16	4.2	0.008
33	19990514	754	34.1	-116.4	4.9	1	19920628	1157	34.20	-116.43	7.6	0.176
							19920423	450	33.96	-116.31	6.3	0.012
							19920915	847	34.06	-116.36	5.6	0.003
43	20000229	1716	28.2	57.1	4.5	1	19990304	538	28.34	57.19	6.6	0.030
48	20000510	2325	33.2	-115.6	3.9	0	19900621	1047	33.16	-115.63	4.0	0.000

Note: Lat: Latitude. Lon: Longitude. M: Magnitude. Acc: accuracy (1:hit, 0: miss). Af Prob. Aftershock probability

Table 14 Possibility of an earthquake in a new moon and a full moon

No	Date	Lat.	Lon.	Mag	Lunar	Y/N	No	Date	Lat.	Lon.	Mag	Lunar	Y/N
1	17690728	34.00	-118.00	6.0	6m24	0	87	19271104	34.70	-120.80	7.3	10M11	0
2	18001122	33.00	-117.30	6.5	10M6	0	88	19320606	40.75	-124.50	6.4	5M3	1
3	18080624	37.80	-122.50	6.0	5m1	1	89	19321221	38.75	-118.00	7.2	11m24	0
4	18121208	34.37	-117.65	7.0	11m5	0	90	19330311	33.62	-117.97	6.3	2M16	1
5	18121221	34.20	-119.90	7.0	11m18	1	91	19330625	39.07	-119.33	6.1	5M'3	1
6	18360610	37.80	-122.20	6.8	4m26	0	92	19340130	38.30	-118.40	6.3	12M16	1
7	183806xx	37.60	-122.40	7.0	*****	x	93	19340608	36.00	-120.50	6.0	4M27	0
8	18521129	32.50	-115.00	6.5	10m18	1	94	19340706	41.25	-125.75	6.0	5M25	0
9	18550711	34.10	-118.10	6.0	5M28	1	95	19341230	32.25	-115.50	6.5	11m24	0
10	18570109	35.70	-120.30	8.3	12M14	1	96	19341231	32.00	-114.75	7.0	11m25	0
11	18570903	39.30	-120.00	6.3	7m15	1	97	19370903	33.40	-116.27	6.0	2m13	1
12	18581126	37.50	-121.90	6.3	10M21	1	98	19400208	39.75	-121.25	6.0	1M1	1
13	18581216	34.00	-117.50	6.0	11M12	1	99	19400519	32.73	-115.50	7.1	4M13	1
14	18600315	39.50	-119.50	6.5	2m23	0	100	19400209	40.70	-125.40	6.6	1M14	1
15	18620527	32.70	-117.20	6.0	4m29	1	101	19410513	40.30	-126.40	6.0	4M18	1
16	18640226	37.10	-121.70	6.0	1m19	0	102	19410914	37.57	-118.73	6.0	7m23	0
17	18651003	37.00	-122.00	6.5	8M19	0	103	19411003	40.40	-124.80	6.4	8M13	1
18	18660715	37.50	-121.30	6.0	6m4	1	104	19421021	33.05	-116.08	6.5	9m12	1
19	18680530	39.30	-119.70	6.0	4m'9	0	105	19450519	40.40	-126.90	6.2	4m8	0
20	18681021	37.70	-122.10	7.0	9m6	0	106	19450928	41.90	-126.70	6.0	8M23	0
21	18691227	39.40	-119.70	6.3	11M25	0	107	19460315	35.73	-118.05	6.3	2m12	1
22	18691227	39.10	-119.80	6.0	11M25	x	108	19470410	34.98	-116.55	6.4	2m19	0
23	18700217	37.20	-122.10	6.0	1M18	1	109	19481204	33.93	-116.38	6.5	11m4	1
24	18710302	40.40	-124.20	6.0	1M12	1	110	19481229	39.55	-120.08	6.0	11m29	1
25	18720325	36.70	-118.10	7.6	2M18	1	111	19490209	41.30	-126.00	6.2	2m26	0
26	18720326	36.90	-118.20	6.8	2M18	x	112	19511008	40.25	-124.50	6.0	9m8	0
27	18720403	37.00	-118.20	6.3	2M26	0	113	19520721	35.00	-119.02	7.7	5M'30	1
28	18720411	37.50	-118.50	6.8	3m4	1	114	19520721	35.00	-119.00	6.4	5M'30	x
29	18721112	39.00	-117.00	6.0	10M12	1	115	19520723	35.37	-118.58	6.1	6m2	1
30	18731123	42.00	-124.00	6.8	10M4	1	116	19520729	35.38	-118.85	6.1	6m8	0
31	18750124	40.70	-120.50	6.0	12m17	1	117	19521122	35.73	-121.20	6.0	10M6	0
32	18751115	32.50	-115.50	6.3	10M18	1	118	19540319	33.28	-116.18	6.2	2m15	1
33	18780509	40.10	-124.00	6.0	4M8	0	119	19540706	39.42	-118.53	6.6	6M7	0
34	18810410	37.40	-121.40	6.0	3m12	1	120	19540706	39.30	-118.50	6.4	6M7	x
35	18830905	34.20	-119.90	6.3	8M5	0	121	19540824	39.58	-118.45	6.8	7m26	0
36	18840326	37.10	-122.20	6.0	2m29	1	122	19540831	39.50	-118.50	6.3	8M4	1
37	18850412	36.40	-121.00	6.3	2m27	1	123	19541024	31.50	-116.00	6.0	9M28	1
38	18870603	39.20	-119.80	6.5	4m12	1	124	19541112	31.50	-116.00	6.3	10m17	1
39	18880429	39.70	-120.70	6.0	3M19	0	125	19541215	40.27	-125.63	6.5	12M1	1
40	18890519	38.00	-121.90	6.0	4M20	0	126	19541216	39.32	-118.20	7.1	11M22	0
41	18890620	40.50	-120.70	6.0	5m22	0	127	19541216	39.50	-118.00	6.8	11M22	x
42	18900209	33.40	-116.30	6.5	1m20	0	128	19541221	40.93	-123.78	6.6	11M27	0
43	18900424	36.90	-121.60	6.3	3M6	0	129	19560209	31.75	-115.92	6.8	12M28	1
44	18900726	40.50	-124.20	6.0	6M10	0	130	19560209	31.75	-115.92	6.1	12M28	x
45	18910730	32.00	-115.00	6.0	6M23	0	131	19560214	31.50	-115.50	6.3	1m3	1
46	18920224	32.55	-115.63	7.0	1m26	0	132	19560215	31.50	-115.50	6.4	1m4	1
47	18920419	38.40	-122.00	6.5	3M23	0	133	19561011	40.67	-125.77	6.0	9M8	0
48	18920421	38.50	-121.90	6.0	3M25	0	134	19561213	31.00	-115.00	6.0	11M12	1
49	18920528	33.20	-116.20	6.5	5m3	1	135	19590323	39.60	-118.02	6.3	2M15	1
50	18940730	34.30	-117.60	6.0	6m28	1	136	19590623	39.08	-118.82	6.1	5M18	1
51	18940930	40.30	-123.70	6.0	9M2	1	137	19600809	40.32	-127.07	6.2	6m17	1
52	18960817	36.00	-118.30	6.0	7m9	0	138	19660628	36.00	-120.50	6.0	5m10	0
53	18970620	37.00	-121.50	6.3	5M21	0	139	19660807	31.80	-114.50	6.3	6m21	0
54	18980331	38.20	-122.40	6.5	3M10	0	140	19660912	39.42	-120.15	6.0	7M28	1
55	18980415	39.20	-123.80	6.5	3M25	0	141	19680415	33.18	-116.13	6.5	3m25	0
56	18990416	41.00	-126.00	7.0	3M7	0	142	19710209	34.42	-118.40	6.5	1m14	1
57	18991225	33.80	-117.00	6.4	11m23	0	143	19761126	41.30	-125.70	6.3	10M6	0
58	19010303	36.00	-120.50	6.4	1m13	1	144	19791015	32.60	-115.30	6.5	8M25	0
59	19030124	31.50	-115.00	6.6	12M26	0	145	19800525	37.60	-118.83	6.1	4M12	1
60	19060418	37.70	-122.50	8.3	3M25	0	146	19800527	37.48	-118.80	6.0	4M14	1
61	19060419	32.90	-115.50	6.2	3M26	0	147	19800609	32.20	-115.08	6.4	4M27	0
62	19060423	41.00	-124.00	6.4	3M30	1	148	19801108	41.12	-124.67	7.2	10m1	1
63	19081104	36.00	-117.00	6.0	10M11	0	149	19830502	33.13	-115.65	6.0	3m22	0
64	19100319	40.00	-125.00	6.0	2M9	0	150	19830502	36.23	-120.32	6.5	3M20	0
65	19100805	42.00	-127.00	6.6	7M1	1	151	19840424	37.32	-121.65	6.1	3M24	0
66	19110701	37.25	-121.75	6.5	6M6	0	152	19860708	40.38	-121.45	6.7	8m15	1
67	19140424	39.50	-119.80	6.0	3m29	1	153	19860708	34.00	-116.60	6.0	6M2	1
68	19150506	40.00	-126.00	6.2	3M23	0	154	19860721	37.53	-118.43	6.2	6M15	1
69	19150623	32.80	-115.50	6.0	5m11	0	155	19871124	33.07	-115.78	6.2	10M4	1
70	19151003	40.50	-117.50	7.3	8M25	0	156	19871124	33.02	-115.85	6.6	10M4	x
71	19151121	32.00	-115.00	7.1	10M15	1	157	19891018	37.04	-121.88	7.1	9m19	0
72	19151231	41.00	-126.00	6.5	11m25	0	158	19910816	41.63	-125.87	6.3	7m7	0
73	19161110	35.50	-116.00	6.1	10m15	1	159	19910817	40.28	-124.23	6.2	7m8	0
74	19180421	33.80	-117.00	6.9	3m11	0	160	19910817	41.68	-126.05	7.1	7m8	x
75	19180715	41.00	-125.00	6.5	6M8	0	161	19920423	33.97	-116.32	6.1	3M21	0
76	19220126	41.00	-126.00	6.0	12M29	1	162	19920425	40.33	-124.23	7.2	3M23	0
77	19220131	41.00	-125.50	7.3	1M4	1	163	19920426	40.43	-124.60	6.5	3M24	0
78	19220310	36.00	-120.50	6.3	2m12	1	164	19920426	40.38	-124.58	6.6	3M24	x
79	19230122	40.50	-124.50	7.2	12M6	0	165	19920628	34.20	-116.43	7.3	5m28	1
80	19230723	34.00	-117.30	6.0	6m10	0	166	19920628	34.20	-116.83	6.2	5m28	x
81	19250604	41.50	-125.00	6.0	4M14	1	167	19930517	37.15	-117.83	6.1	3m'26	0
82	19250629	34.30	-119.80	6.3	5M9	0	168	19940117	34.22	-118.53	6.7	12m6	0
83	19261022	36.62	-122.35	6.1	9m16	1	169	19940901	40.45	-125.90	6.9	7M26	0
84	19261022	36.55	-122.18	6.1	9m16	x	170	19940912	38.82	-119.62	6.0	8m7	0
85	19261210	40.75	-126.00	6.0	11M6	0	171	19950219	40.62	-125.90	6.6	1m20	0
86	19270918	37.50	-118.75	6.0	8M23	0	172	19991016	34.60	-116.27	7.0	9M8	0

Note: In Column "Y/N", "0":86, "1":74, "x":12

Note: Lat. Lon. and Mag. respectively indicate latitude, longitude, and magnitude. In "Lunar": M and m respectively indicate 30-day and 29-day lunar months; "Y/N": 1 and 0 respectively indicate a big earthquake in or not in ±3 days around a new moon or a full moon. "x" indicates no date or an earthquake closely related to another earthquake of the 160 earthquakes. In the total of 160 earthquakes, 1 and 0 contain a rate of 46.2% and 53.8% respectively. The earthquake data were from SCEDC. The lunar data were from "New Almanac (1840~2050)" by the Zijinshan Observatory of the China Scientific Academy (1988).

Appendix

Here are several copies of Shou's earthquake predictions witnessed by the USGS or people, or the both. The examples include hits and misses for big (≥6) ones or moderate (4~5.9). Seismologists and people are all welcome to simulate them: either hits or misses. One can select a series of random dates, or a future date with any method to replace a predicted date to see whether the series or the future date is better than the result of the prediction.

1. The Bam prediction was witnessed by thousands scientists and people. Orhan Cerit (**a**), I-Wan Chen (**b**), and Bulent Doruker (**c**) knew my success even earlier than I. The exact time of this prediction was at "24-Dec-2003 16:58" (**d**) to the Public, recorded by automatic time recorder in Los Angeles, or 0:58 on Dec.25, 2003 in UTC.

a
From: "Çevre Müh. Bölümü" <muhcevre@cumhuriyet.edu.tr> Add to Address Book

To: zhonghao_shou@yahoo.com

Subject: A great Job !

Date: Fri, 26 Dec 2003 20:16:06 +0200

___

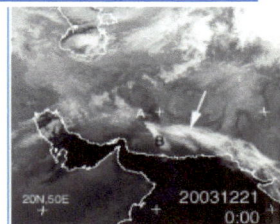

12/21/2003 0:00 | An Iran EQ Cloud, >=5. Likely >=5.5 within 60 days

Your prediction above is a very great job.
Sincerely yours
Orhan Cerit

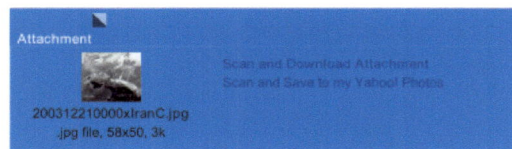

Attachment
Scan and Download Attachment
Scan and Save to my Yahoo Photos
200312210000xIranC.jpg
.jpg file, 58x50, 3k

b From: "Chen I-wan" <cheniwan@263.net> Add to Address Book

To: "Bulent Doruker" <doruker@bnet.net.tr>, "'Quake/LarryBerg'" <larryberg@adelphia.net>, "'Quake/CzPavelKalenda'" <pkalenda@volny.cz>, "'Quake/EDG'" <edgrsprj@ix.netcom.com>, "'Quake/India Shan'" <vu2rss@eth.net>, zhonghao_shou@yahoo.com

Subject: Re: Iran EQ

Date: Fri, 26 Dec 2003 22:57:27 +0800

Dear Mr. Shou Zhong-hai,
Very good!
We must congratulate you for your successful prediction!
An Iran EQ Cloud, >=5. Likely >=5.5 within 60 days
Rgds
Chen I-wan, Advisor
Committee of Natural Hazard Prediction of China Geophysics Society

c —— Original Message ——
From: Bulent Doruker
To: 'Chen I-wan' ; 'Quake/LarryBerg' ; 'Quake/CzPavelKalenda' ; 'Quake/EDG' ; 'Quake/India Shan'
Sent: Friday, December 26, 2003 10:25 PM
Subject: RE: Iran EQ
Hi Chenny and All,
Mr. Zhonghao Shou's prediction based on earthquake clouds (which can be identified as plasma clouds) is as follows.
http://quake.exit.com/images_2003_4.htm

12/21/2003 0:00 | An Iran EQ Cloud, >=5. Likely >=5.5 within 60 days

Rgds,
Bulent Doruker

d

200312210000xIranC.jpg   24-Dec-2003 16:58   4k

200312210000xIranCB.jpg   24-Dec-2003 16:58   30k

2. Shou's first earthquake prediction hit the M7.7 Iran quake on June 20, 1990.

Prediction:
Time: LT 11:45 am (UTC 3:45) of June 20, 1990~
Area: the source of the cloud, WNW far from
Hangzhou (30.3, 120.2)
Size: like the Tangshan earthquake

Witnesses:
Wu, Jiang, "That day, Zhonghao Shou pointed toward that cloud. Next day, news reported an earthquake" on June 23, 1990.
Chen, Jin-Xing, "This prediction is true" on June 23, 1990."

Earthquake: 19900620 21:00 36.96 49.41 7.7 Iran

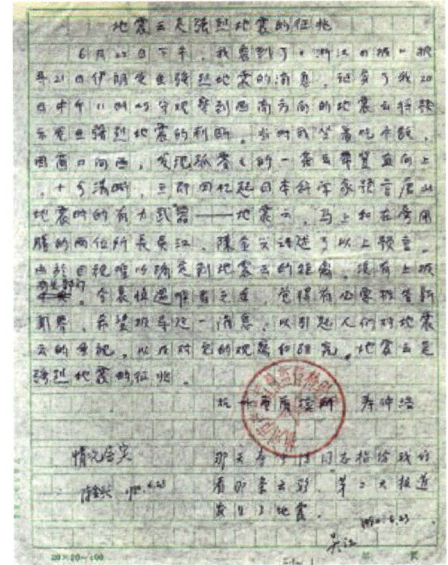

3. No.17 (left) prediction, verified by the USGS on Mar.6, 1997 (LT), hit the M6 China quake on Apr.6 (UTC).
4. No.24 (right), verified by the USGS on Jan.5, 1998, hit the M6.1 Afghanistan quake on Feb.4.

3/6/1997   Pasadena

According to my analysis, I predict that there will be an earthquake of magnitude ≥ 6 M$_L$ in Northern China within 30 days. The most likely place is latitude > 35° and longitude 105~115°

Zhonghao Shou

3/6/1997

Received on
03/06/97 @ 2:25p
by Linda Curtis
U. S. Geological Survey

1/5/1998 Pasadena

I predict that there will be an earthquake of magnitude ≥ 6 on the Richter Scale in 25~41N and 53~105E within 44 days. (A random prediction of equal precision on that area has a 28.6% chance of being correct).
A more precise estimate is: within 30 days, 30~37 N, 58~95 E. (A random prediction of equal precision on that area has a 13% chance of being correct). The area is in Pamir, which consists of parts of Pakistan, Afchanistan, Tajikistan, and China.

Zhonghao Shou

Zhonghao Shou

01-05-98   10:10 AM
Recieved by
Linda Curtis
U. S. Geological Survey
@ Caltech

5. No.31(left), verified by USGS on Feb.22, 1999, hit the M6.6 Iran quake on Mar.4.

6. No.51(right), verified by USGS on Jun.29, 2000, hit the M6.2 Japan quake on Jul.1.

2/22/1999 Pasadena

According to my analysis, I predict that there will be an earthquake of magnitude ≥ 5.5 ML in Asia (20~38N, 50~100E) within 45 days.

The most likely magnitude is ≥ 6 ML.

More likely latitude is 28~38 N.

The most likely latitude is 29~35 N.

More likely longitude is 65~74 E.

The most likely longitude is 68~72 E.

More likely time is 3/1 ~ 4/1.

The most likely time is 3/5 ~ 3/30.

Zhonghao Shou

Received 2/22/99 Lucy from USGS

Zhonghao Shou, 6/29/00 10:35 AM -0700, New Prediction

Date: Thu, 29 Jun 2000 10:35:55 -0700 (PDT)
From: Zhonghao Shou <zhonghao_shou@yahoo.com>
Subject: New Prediction
To: Linda Curtis <linda@usgs.gov>

Dear Linda Curtis, I have to see the dentist now. So, I write a new prediction to you. 6/29/2000 Pasadena According to my analysis, I predict that there will be an earthquake of magnitude >= 6 in Japan (<37N) within 52 days. More likely time is 6/29-7/25. The most likely time is 6/29-7/10. More likely magnitude is >=6.5 ML. The most likely magnitude is >=7 ML. More likely latitude is 33-36 N. The most likely latitude is 34-35.2 N. More likely longitude is 138- 140.5 E. The most likely longitude is 139.1-139.7 E.
Zhonghao Shou

Zhonghao Shou

"Earthquake Clouds and Short Term Prediction"
http://members.xoom.com/EQPrediction/

Received by e-mail on 06/29/00 by Linda Curtis of USGS @ Caltech

7. No.52 (left), verified by USGS on Jul.5, 2000, hit the M6.5 Japan quake on Jul.30.

8. D4 (right), verified by USGS, and witnessed by the Public on May 17, 1999, hit the M7 Mexico quake on Jun.15. ABC TV reported the success on Jun.25.

7/5/2000 Pasadena

According to my analysis, I predict that there will be an earthquake of magnitude >= 6 ML in Japan or the East Chinese Sea (<34N, <142.5E) within 47 days.

More likely time is 7/5~8/5.

The most likely time is 7/5~7/25.

More likely magnitude is >= 6.3 ML.

The most likely magnitude is >= 6.6 ML.

More likely latitude is 25~33 N.

The most likely latitude is 29~32 N.

More likely longitude is 125~135 E.

The most likely longitude is 130~132.5 E.

Zhonghao Shou

Zhonghao Shou

5/17/1999 Pasadeda

According to my analysis, I predict that there will be an earthquake of magnitude ≥ 5 ML in Mexico (< 29N) within 48 days.

More likely time is 5/20~6/20.

The most likely time is 5/25~6/15.

More likely size is ≥ 5.5 ML.

The most likely size is ≥ 6 ML.

More likely latitude is 17~27 N.

The most likely latitude is 22~25 N.

More likely longitude is 100~109 W.

The most likely longitude is 104~106W.

Zhonghao Shou

Zhonghao Shou

Received by Linda Curtis of USGS on July 05, 2000 @ 3:45p

Received by Linda Curtis of U.S. Geological Survey on 05/17/99 2:55p

9. No.57, verified by USGS on Apr.3, 2001, hit an M4.1 Hollister, California quake on Jul.2 exactly.

4/3/2001 Pasadena
According to my analysis, I predict that there will be an earthquake of magnitude >=4 in the white cut area of Image 20010320 15:30, California with an error 30 km within 90 days.
More likely time is 4/3 ~ 6/3.
The most likely time is 4/10~5/10.
More likely magnitude is one 4.2~5.2 ML.
The most likely magnitude is one 4.3~4.8 ML
More likely area is in the cut place.
The most likely area is at "E" in the cut place.

Zhonghao Shou

Zhonghao Shou

Note: This is just an experiment with a big risk for study. The predicted earthquake has not been reported by the World Earthquake Database of the USGS yet.

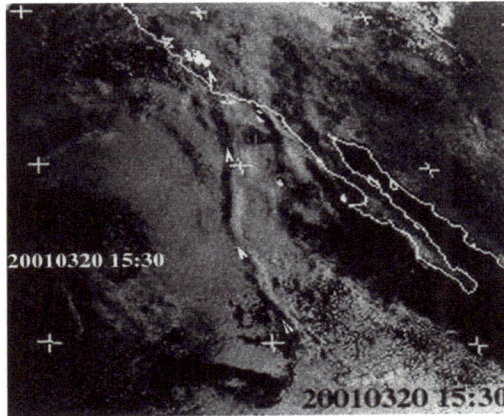

20010320 15:30
20010320 15:30

10. The left prediction to the Public on Dec.14, 2004 hit the M9 Indonesia tsunami earthquake on Dec.26 coincidently.

11. The right prediction at LT 18:25 Oct.6, 2005 or UTC 0:25 Oct.7 hit the M7.6 Pakistan earthquake on Oct.8 correctly. All predictions to the Public adopt a unique time window of 49 days (1999), 103 days (2004), 105 days (early 2005), and 112 days (later 2005).

| 12/13/2004 6:00 | A W Indonesia | | 9/28/2005 8:00 | China or Neighbor EQ Cloud, | |
| (12/15 0:23) | Geoeruption, >=5 | | (10/7 0:25) | >=5, likely >=5.5 | |

200412130600xvIndone..> 14-Dec-2004 18:23   5k   200509280800xWChinaC..> 06-Oct-2005 18:25   4k
200412130600xvIndone..> 14-Dec-2004 18:23   43k   200509280800xWChinaC..> 06-Oct-2005 18:25   30k
200412130600xvIndonesGB.jpg (JPEG Image, 262 × 216 pixels)   200509280800xWChinaCL.jpg (JPEG Image, 227 × 223 pixels)

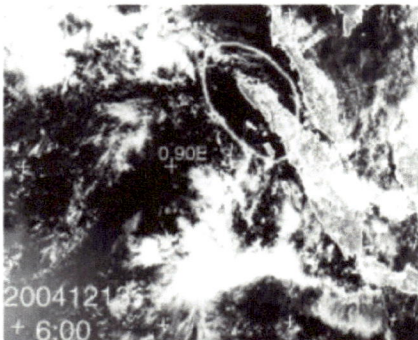

LT "14-Dec-2004 18:23" UTC 0:23 Dec 15   LT "06-Oct-2005 18:25" UTC 0:25 Oct 7

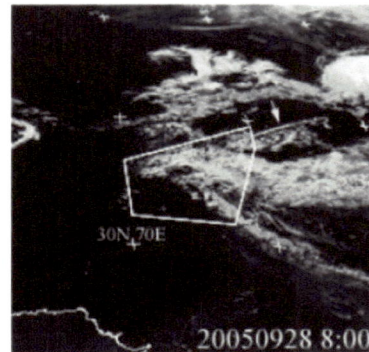

12. A prediction to the Public on Aug. 10, 1999 claimed next major quake not in Los Angeles, predicted by a team of American scientists on Aug. 3, but in two hot pleases: the black triangle **ABC** including Palm Springs, Landers, and part bounder **cd** between California and Nevada (Fig.32b). The M7.4 Hector Mine quake near Landers proclaimed its success on Oct.16, 1999.

EARTHQUAKE CLOUDS AND SHORT TERM PREDICTION

California Earthquake Situation Analysis

- August 10, 1999 -

Many people ask me whether places they live would be safe in an earthquake or not. I would like to tell them if they are in a safe place, but to warn them against a dangerous area, I write this essay to point out what places are relatively dangerous in order to give people time for preparedness. I hope that people living in the relatively dangerous area will not feel panic, but pay more attention to any impending danger.

A satellite image at 12:00 noon (Greenwich time) of July 26, 1999 (1) showed that there was a black triangle in a white cloud. Under the black triangle was a geothermal region, covering a part of three areas: Southern California, Western Nevada, and Northern Mexico. Is it a high coincidence? Furthermore, it looks just like the black hole of the 6.1 Afghanistan earthquake cloud on our cover page. Is it another high coincidence? How did they form?

A meteorologist from UCLA, whose field was special clouds, told me that the meteorology theory could explain neither how the hole of the 6.1 Afghanistan earthquake cloud formed, nor how the small white line-shaped cloud or the 6.1 earthquake cloud formed in the hole. While my earthquake prediction model can. According to my model, there is a hot region around a big impending hypocenter, from which geothermal energy conducts upward, and sometimes can reach to the surface before an earthquake occurs. As the thermal energy reaches to the surface, it heats up the air. The warmed air convects upward, and when it reaches to a weather cloud, a hole forms. The black hole is a clue, implying that there is an impending hypocenter there. This view was supported by my successful Afghanistan earthquake prediction on January 5, 1998.

A difference between this black triangle and that black hole is that the black triangle does not have a line-shaped cloud inside; this means that an earthquake may not come within 49 days.

On July 21, 26, 27, and 28, many earthquake clouds appeared. I photographed them, but not at the times of their initial formation. Moreover, winds confused me about where those clouds came from. The 5.6 mb California-Nevada border earthquake on August 1 was related to the earthquake clouds on July 21 which many people probably saw. But where other clouds came from is still a puzzle, and we need pay more attention to the 49 days since July 26.

Although the 5.6 California-Nevada border earthquake released part of thermal energy, the problem has not disappeared. On the evening of August 3, TV Channel 7 reported temperature of 109 degree at Palm Spring as the highest in Southern California, and the weather reporter said that he could not understand why that place was so hot. On August 4, he reported 104, the highest temperature, at Palm Spring, 90 at the northwest, 94 at the west, 84 at the southwest, and lower than 80 at other places (except both the east and southeast because of no data). This temperature distribution is harmonious with the color-relief (the darker a place, the higher its temperature) of recent satellite images (2).

Earthquakes (>=4) in 1999 are active in three regions (3). One is the black triangle , containing **Palm Springs, Landers, Imperial Valley, Volcano Lake (Mexico), and so on** . Another is the border between **California and Nevada** (Latitude > 35 N, longitude < 119 W). The other is the off coast of Northern California. This distribution is compatible with the color-relief map of satellite images (1, 2). According to the color- relief and earthquake data, I think that Los Angeles, San Francisco, San Diego, Parkfield, and Northridge will be safe this year. The coastal region is less active than the others. The black triangle indicates a dangerous place.

Since I lack research resources, and there are not enough detailed satellite images for earthquake prediction, I may miss some earthquake clouds, or be unable to determine where they came from. To help people protect themselves against big earthquakes, I propose a few photographs of important earthquake clouds (4~6). You will be able to check a cloud with those photographs. If you are rich enough, you can set up an automatic video camera to scan the sky. This will be helpful for your safety. To detect the time, I also propose other precursors such as gas, water, or oil eruption, sulfurous odor, earth noise, sudden gaps, and earth fire called "earth light".

Recently, geophysicists at **NASA's Jet Propulsion Laboratory (JPL) issued a warning that "L.A.'s Big Squeeze likely site of next major quake"**. They predicted that, **"the heart of the city will be struck"** (7). That is a good attempt.

Many seismologists insist that predicting earthquakes is impossible. They deny my work, but have not attempted to respond to my two "Yes or No" questions: Can they explain how the 6.1 Afghanistan earthquake cloud on our cover page formed? Can they make some earthquake predictions as good as mine? They should do better than me because of their rich sources and foundations.

I respect the attempt of scientists of JPL, but **do not think that next big earthquake will attack the city of Los Angeles**. According to my analysis, **the next big one will be in either the black triangle, or the California-Nevada border**. To detect where is the most dangerous place, I suggest residents in Southern California, Northern Mexico (latitude > 28 N), and Nevada (longitude > 110 W) measure outside temperature with a thermometer or a thermograph once a day between 2 and 4 a.m. for ten days, then e-mail me your data with the latitude and the longitude where the date were obtained. This action may help both me and yourselves to find the hottest or the most dangerous place roughly. In fact, the best way to figure out the most dangerous place is to make a grid net, having a distance 10 km between observation points, and use the net to measure temperatures at three depths: surface, a depth of about 0.8 meter, and 1.6 meter under the ground once a day at 2 or 4 a.m..

Finally, I should thank many people for believing my work, telling me of a web site for surface wind distribution, and offerng me their best wishes. I thank the USGS for earthquake data, thank Dundee University, UK and its web master Andrew Brooks, and Utah University for satellite images.

References

1. *Martin Kasindorf. L.A.'s Big Squeeze likely site of next major quake. USA TODAY. 8/3 3A (1999).

Home | Introduction | Publication & News | Predictions | New Predictions | Essays
| Links | Contact

Sign Our Guestbook 🔲 ⌒ View Our Guestbook

Updated: August 10, 1999 | Webmaster

References

Abbott, A. & Nosengo, N. (2014). Italian seismologists cleared of manslaughter. *Nature* **515,** 171.

Ahrens, C.D. (1991). *Meteorology Today.* (West Publishing Company, St. Paul, MN).

Bolt, B.A. (1988). Earthquakes. Revised and Updated (W.H. Freeman).

Bowen, N. & Aurousseau, M. (1923). Fusion of sedimentary rocks in drill-holes. *BullGeol Soc Am* **34 No. 3,** 431–438.

Cai, Y. G., Yin, Y.Q. & Wang, R. (1987) A study on preseismic fault creep and ground temprature anomaly. *Acta Seismologica Sinica* **9, No.2,** 167-175.

Campbell, W. H. (2006) A misuse of public funds: U.N. support for geomagnetic forecasting of earthquakes and meteorological disasters *EOS* DOI: 10.1029/98EO00354

China Academy of Building Research (1986). The Mammoth Tangshan Earthquake of 1976 Building Damage Photo Album (Beijing: China Academic Publishers).

Clarke, Tom (2001). Water thrown on earthquake prediction. *Nature* **412,** 812 - 815

Cox, A., & Doell, R. R. (1962). Magnetic properties of the basalt in hole EM 7, Mohole project. *Journal of Geophysical Research*, **67(10),** 3997-4004.

Cox, K.G. (1978). Kimberlite pipes. *Sci. Am* **238,** 4.

Geller. R (1996). Debate on evaluation of the VAN method: Editor's introduction. *GRL* **23-11** 1291-1293.

Geller, R.J., Jackson, D.D., Kagan, Y.Y., and Mulargia, F. (1997). Earthquakes Cannot Be Predicted. *Science* **275,** 1616.

Gerstenberger, M. C., Wiemer, S., Jones, L. M., & Reasenberg, P. A. (2005). Real-time forecasts of tomorrow's earthquakes in California. *Nature*, **435**(7040), 328-331.

Haas, J.L.J. (1971). The effect of salinity on the maximum thermal gradient of a hydrothermal system at hydrostatic pressure. Eco Geol 940–946.

Haicheng Earthquake Study Delegation (1977). Prediction of the Haicheng earthquake. Eos Trans. AGU **58,** 236–272.

Hall, S. S. (2011). Scientists on trial: At fault? *Nature* **477,** 264-269.

Harrington, D., and Shou, Z. (2005). Bam Earthquake Prediction & Space Technology. *Seminars of the United Nations Programme on Space Applications* (United Nations) **16.** 39–63.

Hess, H. H. (1962). History of ocean basins. In *Petrologic studies: A volume to honor of A.F.Buddington* (ed. Engel, A.E.J. et. al. GSA, Boulder, CO.) **28,** 599-620.

Holliday, J.R., Gaggioli, W.J., and Knox, L.E. (2012). Testing earthquake forecasts using reliability diagrams. *Geophys. J. Int.* **188,** 336–342.

Holmes, A. (1931). XVIII. Radioactivity and Earth Movements. *Transactions of the Geological Society of Glasgow*, **18** (**3**), 559-606.

Hopkin, M. (2004). Sumatran quake sped up Earth's rotation. *Nature*. **30**: 041229-6.

Hough, S. E. (2002). *Earthshaking science: what we know (and don't know) about earthquakes*. Princeton University Press.

Huang, S., Jean, J., and Hu, J. (2003). Huge rock eruption caused by the 1999 Chi-Chi earthquake in Taiwan RID E-7509-2010. *Geophys. Res. Lett.* **30**, 1858–1862.

Isacks, B., Oliver, J., & Sykes, L. R. (1968). Seismology and the new global tectonics. *Journal of Geophysical Research,* **73**(**18**), 5855-5899.

Jinchang, J., & Zhang, D. (1984). A study on the relationship between the events of hibernating snakes crawling out from their holes (EHSCH) and the earthquakes. *Journal of Seismological Research* **6**, 012.

Jones, R.H., and Jones, A. (2003). Testing skill in earthquake predictions. *Seismological Research Letters* **74**, 753–759.

Kanamori, H. (1983). Magnitude Scale and Quantification of Earthquakes. *Tectonophysiscs* **93**, 185-199

Karakelian, D. Klemperer, S. L. Fraser-Smith, A. C. and Beroza, G. C. (2000) A Transportable System for Monitoring Ultra Low Frequency Electromagnetic Signals Associated with Earthquakes. *SRL* **71-4**, 423-436.

Killick, A. (1990). PSEUDOTACHYLITE GENERATED AS A RESULT OF A DRILLING BURN-IN. *Tectonophysics* **171**, 221–227.

Kirby, S., and McCormick,, J.. (1990). Practical Handbook of Physical Properties of Rocks & Minerals (Florida: CRC-Press).

Koch, N., and Masch, L. (1992). Formation of Alpine mylonites and pseudotachylytes at the base of the Silvretta nappe, Eastern Alps. *Tectonophysics* **204**, 289–306.

Kossobokov, V.., Romashkova, L.., Keilis-Borok, V.., and Healy, J.. (1999). Testing earthquake prediction algorithms: statistically significant advance prediction of the largest earthquakes in the Circum-Pacific, 1992–1997. *Physics of the Earth and Planetary Interiors* **111**, 187–196.

Lambeck, K. (1980). The Earth's Variable Rotation: Geophysical Causes and Consequences (Cambridge University Press).

Lane, T., and Waag, C. (1985). Ground-water eruptions in the Chilly Buttes area, Central Idaho. *Special Publications*, p. 19.

Langbein, J. (1992). The October 1992 Parkfield, California earthquake prediction *Earthquakes & Volcanoes* **23,**160-169.

Le Pichon, X. (1968). Sea-floor spreading and continental drift. *Journal of Geophysical Research*, **73**(12), 3661-3697.

Li, D.J. (1982) Earthquake Clouds, 148-150 (Xue Lin Public Store, Shanghai, China) in Chinese.

Luen, B., and Stark, P., B. (2008). Testing earthquake predictions. *IMS Collections* **2**, 302–315.

Maddock, R. (1983). MELT ORIGIN OF FAULT-GENERATED PSEUDOTACHYLYTES DEMONSTRATED BY TEXTURES. *Geology* **11**, 105–108.

Maddock, R.H. (1992). Effects of lithology, cataclasis and melting on the composition of fault-generated pseudotachylytes in Lewisian gneiss, Scotland. *Tectonophysics* **204**, 261–278

Magloughlin, J.F. (1992). Microstructural and chemical changes associated with cataclasis and frictional melting at shallow crustal levels: the cataclasite-pseudotachylyte connection. *Tectonophysics* **204**, 243–260.

Marris, Emma. (2005) Inadequate warning system left Asia at the mercy of tsunami. *Nature* **433**, 3-5.

McCalpin, J. P. (1996) Earthquake Magnitude Scales. (ed. McCalpin, James P. Academic Press. 525 B Street, Suite 1900, San Diego, California 92101-4495, USA).

Meyerhoff, H. A. (1972). The New Global Tectonics": Major Inconsistencies. *AAPG Bulletin* **56** No. 2, 269-336.

Molchan, G., and Romashkova, L. (2011). Gambling score in earthquake prediction analysis. *Geophys. J. Int.* **184**, 1445–1454.

Mormile, Dennis (1994) Japan holds firm to shaky science. *Science* **264,** 1656-1658.

Nosengo, Nicola (2012) L'Aquila verdict row grows *Nature* **491**, 15–16

O'Hara, K. (1992). Major- and trace-element constraints on the petrogenesis of a fault-related pseudotachylyte, western Blue Ridge province, North Carolina. *Tectonophysics* **204**, 279–288.

Passchier, G. (1982). Mylonitic deformation in the Saint- Barthelemy Massif, French Pyrenees, with emphasis on the genetic relationship between ultramylonite and pseudotachylite. *GUA Munic. Univ Amst. Pap Geol* **1**, 1–173.

Peltzer, G., Crampé, F., & King, G. (1999). Evidence of nonlinear elasticity of the crust from the Mw7. 6 Manyi (Tibet) earthquake. *Science*, **286**(5438), 272-276.

Peltzer, G., Crampé, F., & Rosen, P. (2001). The Mw 7.1, Hector Mine, California earthquake: surface rupture, surface displacement field, and fault slip solution from ERS SAR data. *Comptes Rendus de l'Académie des Sciences-Series IIA-Earth and Planetary Science*, **333**(9), 545-555.

Quan, Y.-D. (1988). The Haicheng, Liaoning Province, Earthquake of M7.3 of 4 February 1975, in *Earthquake Cases in China (1966–1975)* (ed, Z.-C. Zhang. State Seismological Bureau Publication in Chinese, Seismological Press, Beijing, 1988), pp. 189-210.

Raff, A. D. (1963). Magnetic anomaly over Mohole drill hole EM7. *Journal of Geophysical Research*, **68**(3), 955-956.

Reasenberg, P.A., and Jones, L.M. (1994). Earthquake Aftershocks: Update. Science **265**, 1251–1252.

Sadowski (САДОВСКИЙ), M. A. (1982) Seismic Electromagnetic Precursor (Chinese ed姚家榴Yao Jie Liu 63 FuXing road Beijing Xin Hua publisher 1986)

Shi, H.X., Cai, Z. H., and Gao, M.X. (1980). Anomalous migration of shallow groundwater and gases in the Beijing region and the 1976 Tangshan earthquake. *Acta Seism. Sin.* **2**, 55–64.

Shou, Z. (1999). Earthquake clouds, a reliable precursor. *Sci. Utopya* **64**, 53–57.

Shou, Z. (2006a). Earthquake Vapor, a reliable precursor. In *Earthquake Prediction*, (ed, S. Mukherjee. V.S.P. Intl Science), pp. 21–51.

Shou, Z. (2006b). Precursor of the largest earthquake in the last 40 years. *New Concepts Glob. Tectonics* **41**, 6–15.

Shou, Z., Xia, J., and Shou, W. (2010). Using the earthquake vapour theory to explain the French airbus crash. *Remote Sens. Lett.* **1**, 85–94.

Shou, Z. (2011). Method of precise earthquake prediction and prevention of mysterious air and sea accidents, *United States Patent*: **8068985**.

Sibson, R. (1975). GENERATION OF PSEUDOTACHYLYTE BY ANCIENT SEISMIC FAULTING. *Geophys. J. R. Astron. Soc.* **43**, 775–794.

Silver, P. G. and Wakita, H. (1996). A Search for Earthquake Precursors. *Science* **273**. 77-78

Smyth, C., Yamada, M., Mori, J., and Anonymous (2012). Earthquake forecast enrichment scores. Research in Geophysics (Testo Stampato) **2**, 7–12.

Spray, J. (1987). Artificial generation of pseudotachylyte using friction welding apparatus: simulation of melting on a fault plane. *J of Struct Geol* **9 (1)**, 49–60.

Spray, J. (1992). A physical basis for the frictional melting of some rock-forming minerals, *Tectonophysics* **204**, 205–221.

Swanson, M. (1992). Fault structure, wear mechanisms and rupture processes in pseudotachylyte generation.. *Tectonophysics* **204**, 223–242.

Techmer, K., Ahrendt, H., and Weber, K. (1992). The development of pseudotachylyte in the Ivrea - Verbano zone of the Italian Alps. *Tectonophysics* **204**, 307–322.

Thatcher, W. (1992). Scientific goals of Parkfield earthquake prediction experiment *Earthquakes & Volcanoes* **20**, 75-82.

The Institute of Geology, State Seismological Bureau (1981). *The photo album of eight strong earthquake disasters in China.* (63 Fuxing Rord Beijing)

Tuefel, L., and Logan, J.. (1978). Effect of displacement rate on the real area on contact and temperatures generated during frictional sliding of Tennessee Sandstone. *Pure Appl. Geophys.* **116**, 840–865.

Utsu, T. (2002). "Statistical features of seismicity." *International Geophysics Series* **81.A**: 719-732.

Varotsos, P. and Lazaridou, M. (1991). Latest aspects of earthquake prediction in Greece based on seismic electric signals. *Tectonophysics* **188.3**: 321-347.

Vine, F. J. and Matthews, D. H. (1963). Magnetic anomalies over oceanic ridges. *Nature.* **199**. 947-949.

Vine, F. J. (1966). Spreading of the ocean floor: new evidence. *Science*, **154(3755)**, 1405-1415.

Wang, C. L., Yao, Q. C., & Long, M. (1980). On the abnormal characteristics of radon contents in the spring waters north of the epicentral area of the Sungpan-Pingwu earthquake before its occurrence and their cause. *Acta Seismol. Sin*, **2**, 83.

Wegener, A. (Verlag, S. 2002 trans.) The origins of continents. *Geol Rundsch* **3**, 276-292.

Wenk, H., and Weiss, L. (1982). A1-RICH CALCIC PYROXENE IN PSEUDOTACHYLITE - AN INDICATOR OF HIGH-PRESSURE AND HIGH-TEMPERATURE. *Tectonophysics* **84**, 329–341.

Winkler, H.G.. (1979). PETROGENESIS METAMORPHIC 4/E, ROCKS 4TH ED SSE RPT (New York: Springer).

Wu, Q.., and Liu, A.. (1983). Anomalous variations in production oil wells before and after the great Haicheng and Tangshan earthquakes. *Acta Seism. Sin.* **5**, 461–466.

Wyss, M (1996) Inaccuracies in seismicity and magnitude data used by Varotsos and coworkers. *GRL* **23-11** 1299-1302

Wyss, M and Allmann, A (1996) Probability of chance correlations of earthquakes with predictions in areas of heterogeneous seismicity rate: the VAN case. *GRL* **23-11** 1307-1310.

Xie, J.M. & Huang, L.R. (1987). The vertical deformation before and after the Tangshan earthquake of 1976. *Seismology and Geology* **9, No. 3**, 1-19.

Yang, C.S. (1982). Temporal and spatial distribution of anomalous ground water changes before the 1975 Haicheng earthquake. *Acta Seism. Sin.* **4**, 84–89.

Zechar, J.D.and Jordan, T.H. (2010). The area skill score statistic for evaluating earthquake predictability experiments. *Pure and Applied Geophysics* **167**, 893–906.

Savage, M.K., Rhoades, D.A., Smith, E.G.C., Gerstenberger, M.C. and Vere-Jones, D. (2010) Introduction. *Pure and Applied Geophysic* in press.

Zhang, D.Y., and Zhao, G.M. (1983). Anomalous variations in oil wells distributed in the Bohai bay oil field before and after the Tangshan earthquake of 1976. *Acta Seism. Sin.* **5**, 360–369.

Zhang Ying-zhen (1981). On the anomalous crustal bulge and aseismic creep prior to the 1976 Tangshan earthquake. *Acta Seismologica Sinica* **3, No.1**, 11-22

Zhuang, J. (2010). Gambling scores for earthquake predictions and forecasts. *Geophysical Journal International* **181**, 382–390.

Zijinshan Observatory of the China Scientific Academy (1988) New Almanac (1840~2050) (ed. Wang, Wing Lu. Scientific Popular Publisher. #32 Bu Zei Jor Road, Hai Den Region, Beijing)

Links

@1 The Southern California Earthquake Data Center (SCEDC) http://www.data.scec.org/ftp/catalogs/SCSN/

@2 The United States Geological Survey (USGS) ftp://hazards.cr.usgs.gov/pde/

@3 The National Climate Data Center (NCDC) http://www.ncdc.noaa.gov/oa/ncdc.html

@4 Dundee University (DU) http://www.sat.dundee.ac.uk/pdus.html

@5 Earthquake Clouds & Short Term Predictions: previous http://quake.exit.com present http://www.earthquakesignals.com/

@6 An animation of the Bam cloud https://www.youtube.com/watch?v=vC-qmbONlxY

@7 An animation of the Indian cloud https://docs.google.com/file/d/0B3PS6mjpf0ITSnN5MHY4NFJncW8/edit?usp=sharing

@8 The Weather Underground (WU) http://www.wunderground.com/

@9 225°C Karachi 20041115 4:30 https://docs.google.com/file/d/0B3PS6mjpf0ITNmhxcUdPNHI2TjQ/edit?usp=sharing

@10 146°C Lahore 20041115 16:30 https://docs.google.com/file/d/0B3PS6mjpf0ITNEN6eHllLUFNSG8/edit?usp=sharing

@11 288°C Karachi 20041116 13:00 https://docs.google.com/file/d/0B3PS6mjpf0ITTlhCV3oxNnRncmM/edit?usp=sharing

@12 69°C NOAA limitation http://www.oso.noaa.gov/goes/goes-calibration/G12_Img_Ch2_Rollover/G12_Ch2_Rollover_Abs.pdf

@13 The Central Weather Bureau of Taiwan (CWBT) http://www.cwb.gov.tw/V7/index.htm

@14 The National Ocean and Atmosphere Administration (NOAA) http://www.goes.noaa.gov/

@15 Kochi University http://weather.is.kochi-u.ac.jp/archive-e.html

@16 The University College London (UCL) ftp://weather.cs.ucl.ac.uk/Weather/

@17 The UN Office for the Coordination of Humanitarian Affairs http://wwwnotes.reliefweb.int/websites/rwdomino.nsf/069fd6a1ac64ae63c125671c002f7289/60adec26e8c12cdec12565c500395fba?OpenDocument

@18 The Space Science & Engineering Center of Univ. of Wisconsin-Madison (SSEC) http://www.ssec.wisc.edu/

@19 The European Organisation for the Exploitation of Meteorological Satellites (EUMETSAT) http://www.eumetsat.int/website/home/index.html

@20 National Climatic Data Center (NCDC) http://www.ncdc.noaa.gov/gibbs/year

@21 Google map http://maps.google.com/

@22 USGS http://pubs.usgs.gov/gip/dynamic/historical.html#anchor9588978

@23 Wikipedia http://en.wikipedia.org/wiki/Continental_drift

@24 European Geosciences Union http://www.egu.eu/egs/meinesz.htm

@25 Wikipedia http://en.wikipedia.org/wiki/Vladimir_Belousov

@26 Meyerhoof,A.A. and Meyerhoof, H.A. http://archives.datapages.com/data/bulletns/1971-73/data/pg/0056/0002/0250/0269.htm

@27 Wikipedia http://en.wikipedia.org/wiki/Sakhalin-I

@28 Wikipedia http://en.wikipedia.org/wiki/Vine–Matthews–Morley_hypothesis

@29 USGS http://geomaps.wr.usgs.gov/parks/deform/gfaults.html